P l a

ROCK
DOC

Planet
ROCK
DOC

Nuggets from Explorations of the Natural World

Dr. Elsa Kirsten Peters

Washington State University Press
Pullman, Washington

Washington State University Press
PO Box 645910
Pullman, Washington 99164-5910
Phone: 800-354-7360
Fax: 509-335-8568
Email: wsupress@wsu.edu
Website: wsupress.wsu.edu

Library of Congress Cataloging-in-Publication Data

Peters, Elsa Kirsten, 1960-
  Planet Rock Doc : nuggets from explorations of the natural world / Elsa Kirsten Peters.
    p. cm.
  Collection of the nationally syndicated Rock Doc newspaper colunms.
  Includes bibliographical references.
  ISBN 978-0-87422-310-1 (alk. paper)
1. Science--Miscellanea. 2. Engineering--Miscellanea. I. Title.
  Q173.P47 2012
  500--dc23

                                                        2012000171

Illustrations by Caryn L. Lawton

Fine Quality Books from the Pacific Northwest

Dedication

For Steve McClure
who gave me the chance to work in a newsroom

and

For Craig Clohessy
who labored tirelessly each day for three years
teaching me how to write for newspapers.

Table of Contents

Acknowledgments

It's a pleasure to acknowledge several people who have helped make possible the Rock Doc column, blog, and radio broadcasts, as well as those who have helped to expand the newspaper arm of the publications effort to reach a national audience with hundreds of thousands of readers from coast to coast.

Steve McClure was the managing editor of the *Moscow-Pullman Daily News* when he allowed me to start running a column about the rocks and fossils found near my hometown. Steve may have been a tad skeptical about how many people would be interested in a column focused on geology, but I give him credit for immediately amending his views when the columns started coming out and readers were interested in them. I've appreciated his support ever since those early days, and his editorial input was a key to the flow of those original columns.

At first the Rock Doc pieces focused solely on geology, but I soon expanded into topics related to energy and other realms in science and engineering. Michael Griswold, when he was dean of the College of Sciences at Washington State University, gave me the time to pursue the goal of expanding the Rock Doc subscription base well beyond the local newspaper. My indefatigable assistant, Susan Bentjen, took on the project of writing to all the newspapers in the Pacific Northwest, shopping around the Rock Doc column. Susan's good work was rewarded with adoptions in both daily and weekly papers, and our subscription base started to grow substantially, soon passing 100,000 readers. For a time Susan was assisted by Elaine O'Fallon in the effort to contact newspaper editors with cold-call letters. Elaine helped us out of the goodness of her heart, not because her work responsibilities included laboring for the Rock Doc effort, and her generosity and goodwill are remembered fondly by both Susan and me.

In a steadily expanding effort, Susan wrote to newspapers in California, other parts of the West, and the Great Plains. Without so much as pausing for breath, she jumped the Mississippi and went on to newspapers in the Midwest, then the Northeast, and finally the Southeast. Due to Susan's dedication and industry, over a period of two years the subscription base grew to include many hundreds of thousands of newspaper readers from sea to shining sea. We knew we had reached a plateau in our need to publicize the Rock Doc column to newspapers when editors started to come on their own to our website (rockdoc.wsu.edu) and use content posted there—as all papers are always welcome to do at any time as long as they credit Washington State University and the College of Agricultural, Human, and Natural Resource Sciences.

In recent months I've been recording audio versions of the Rock Doc columns for Northwest Public Radio. The only reason my readings are useable for broadcast is that NWPR announcer Robin Rilette has been sitting beside me in the recording booth, patiently coaching me at every turn. I greatly appreciate a chance—at this advanced age—to try to master a new skill.

Our next step in the total, wall-to-wall Rock Doc effort will see Susan Bentjen posting the NWPR audio recordings to our website and starting the process of soliciting other public radio stations around the nation to broadcast the audio files. Although I'll never be a famous radio announcer, maybe I'll be heard on a radio station near you in the coming years.

Public relations professional Barbara Petura of WSU was an early friend of the Rock Doc effort. She not only edited scores and scores of columns, she plugged Susan and me into social media and made suggestions to improve our website. While some public relations professionals would have been so concerned about the Rock Doc's clear independence and strong temperament that they would not have collaborated with the effort, Barb met me more than half way, time and time again.

In recent years I've been working at WSU for Ralph Cavalieri, director of the Agricultural Research Center. Ralph has allowed me the time to continue column publication efforts, as well as pursue the radio broadcast work. I deeply appreciate Ralph's support. Kathy Barnard and others in her team of communications professionals now routinely help edit the column and provide other support to the ongoing effort. Thanks to them all.

Finally, Mary Read and others at the WSU Press have been crucial to the process of envisioning and producing this book. The editorial and typography contributions of Kerry Darnall and Nancy Grunewald bring a fresh presentation to this collection, and Caryn Lawton's original pen and ink drawings beautifully illustrate each section. My favorite project in all the world is always to work on a book, and I deeply appreciate publishers who help bring my manuscripts to readers.

To reiterate what I hope is obvious, the Rock Doc has come a long way since its beginnings in a single newspaper with a subscription base of 7,000. I know that none of this progress would have occurred without the fierce productivity Susan Bentjen brings to every task she undertakes. Greater assistants are there none, and I am truly in Susan's debt.

—*Dr. Elsa Kirsten Peters, the Rock Doc*

Foreword

I know Kirsten Peters well. The professional article based on the research she did with me as an undergraduate at Princeton some 30 years ago is my most cited article; she, another undergraduate, and two graduate students were coauthors. She went on to do a superb job on her doctorate at Harvard. Unfortunately, as she alludes to in several pieces in this book, there aren't enough jobs at the cutting edges of science to support all the doctorates that the United States produces. Yes, there are good jobs in industry, but this was not satisfying to Kirsten. She wanted to stay within sight of major science developments, and she does this by writing about these advances for us. There is an adage that "if you cannot explain the science you are working on to your grandmother, then maybe you don't understand it." Kirsten seeks out those scientists who *can* explain their projects to her and then she explains the project to her "grandmother"—that is, to all of us. That leaves scientists to focus on their research, and Kirsten explains to us why our tax dollars are such great investments when applied to basic research.

This book brings together dozens of Kirsten's newspaper columns. Common themes in her life are sprinkled through the pieces and these helped draw me into each story. I found myself identifying with Kirsten even though I don't have arthritis, don't own a gas-guzzling pickup, am not particularly religious, and am not a descendent of Swedish immigrants. On another level, I understand her triumphs at making complex subjects simple enough for a curious, casual reader of the daily newspaper. I have tried to do this myself, and I have kept up with Kirsten's career through our mutual interest in trying to reach an audience beyond a narrow scientific discipline. From my shortcomings in this endeavor, I appreciate all the more what Kirsten has accomplished.

Like Kirsten, I am a geologist; it comes as no surprise that I found her geology columns the easiest to follow. I may not be objective in commenting on them; nevertheless, the one on Iceland is a real tour de force. It not only merges Kirsten's ancestors with the history of democracy and with the most important geologic fact of plates spreading apart at oceanic ridges, but also with contemporary politics and the economics of energy production.

The column on fracking is a must-read for all of us trying to grasp the politics of environmental agendas versus our needs for new energy sources.

In other sections of this book I learned about the economics of food production, the science of tree rings and climate change, and staph infections.

From each of the pieces in this book, I come away with that satisfactory feeling that comes from a new understanding. That is the "high" that Kirsten says she gets when she has completed a piece. There are dozens of potential highs in this book!

—Professor Lincoln S. Hollister
Department of Geosciences
Princeton University

Introduction

I t's sobering but also interesting to reach the point in life when you know there are fewer years ahead of you than lie in your wake. I've reached that place, and reflecting on what I've been fortunate enough to do so far brings to mind a number of different efforts at science outreach and education.

In a variety of venues over the last 15 years, I've introduced non-scientists to topics as varied as the rich beauty of the fossil record, the technical evidence of the Big Bang, recent advances in the mathematics of computerized face-recognition, how scientists know that climate has repeatedly changed in the last 10,000 years, and the all-important changes in food technology that have led to the invention of "cheese bread." Naturally, as a geologist, I've put a disproportionate amount of time into exploring topics like earthquake risks, volcanoes, and my own personal favorite, gold ore and where to find it. Because I've been lucky enough to learn directly from research scientists in a variety of disciplines about current advances in technical knowledge, I've also written about sunlight's interaction with our DNA, the appearance of new disease strains that threaten wheat production, and our prospects for one day splitting water molecules as an energy supply. My subjects have been as eclectic—or perhaps as disorganized—as the cacophony of my daily thoughts. But I hope that woven through the years there has been a unifying thread that simply celebrates the joys of learning about the natural world.

It's not clear why I've put so much time and energy into science outreach. Maybe it's a calling, but I doubt it. I never had a road-to-Damascus experience telling me I should dedicate myself to explaining science in the public square. I simply woke up one day and found myself doing so. Over the years I've used a number of different approaches to address technical life and scientific knowledge, but I think it's not any higher calling, it's just my self-assigned job. Surely it seems to be my place.

In modern society citizens make use of the fruits of science on a daily basis. And every day, through our federal taxes, we all help support research into fields as varied as cancer treatment and developing better stains of wheat. Given how intimately citizens are tied into the many tentacles of science, I think it's important for at least the elementary parts of the subject to be understood by as many people as possible.

Another reason I'm glad I've put a significant amount of time into science outreach is that laboratory and field research is sometimes feared because of the false assumption that it's too difficult for a normal person to understand. The good news, I believe, is that everyone can grasp the basics of science and—most importantly—the direct and simple methods that scientists have used to learn more about the natural world. Scientists look at whatever evidence they can find to help them understand the way Mother Nature works. True, the evidence crucial to advances in science has grown a bit more esoteric over the generations. But my unshakable belief is that there's really nothing terribly difficult in understanding much of science, at least not if a person has a willing guide to it. I try to be such an escort for the public because I believe citizens are better off when both the natural world and scientists themselves are as demystified as they can be. In an era when our national economy and our personal lives are shaped by science and technology, everyone deserves the best chance possible at understanding how empirical work proceeds, how it is funded, and what it's likely to discover next.

But that's the abstract part of my motivation. At a more visceral level, I feel such a deep wellspring of joy when I learn something significant about the world around us, it's natural for me to want to share my pleasure. The internal delight that ideas bring can be a powerful—and fully legal—high. I suspect that joy long ago created in me a positive addiction that leads me to truly want to study technical information in the early mornings before I go to work. Sometimes that means reading heavy tomes during long winters in my cold and clammy basement study, but so be it. I like the rush of learning, and if I can give others even a fragment of the joy I garner from scientific studies, my efforts at public outreach will add to the sum total of happiness in the world.

From time to time I'm also motivated by a desire to fight against some currents in our society that act against a broader knowledge of science. Just as one example, I fear some Americans have been discouraged from learning much about technical life due to a common report that children in America are not being taught sufficient science and math in the public schools to compete in the world economy.

But there's good reason to fight against the drumbeat condemning our young people and their supposedly poor preparation. Comparisons show that while many international students do indeed excel at the drill-and-practice approach to education, they are much more at sea than our young people when it comes to problem-solving or self-directed learning. At its core, science isn't about hammering material into one's head. It's a creative discipline as much as writing verse or forging a constitution, and learning how to regurgitate answers to narrow questions is no preparation at all for what really matters in science, nor indeed in intellectual life more generally.

I work at a land-grant university where most undergraduates have a public school background when they arrive to start their higher educations. Students majoring in science at public universities do good work in all sorts of demanding classes. And they now routinely help with meaningful intellectual work in research laboratories before they graduate. That's right—they work alongside graduate students and faculty members pursuing fundamental and open-ended questions about how the natural world really works. In my day, that could be said only about a scattered few undergraduates across the nation, mostly congregated in elite private institutions. The fact that today undergraduates at public universities are doing real research science deserves much broader recognition, in part because it clearly suggests ordinary mortals can—in fact—comprehend science and participate in the realm of the best research endeavors our society has to offer.

A purely autobiographical fact also suggests that science and math really can be mastered by ordinary Americans. I started out in college as a philosophy major destined, I believed, to go to law school. I switched from that path to the life of a geology major at the last possible moment. Thus I was past my 21st birthday before I began taking serious science or math classes with upperclassmen. Still, I earned a bachelor's degree and went on to earn a doctorate in geology. These facts reflect not that I was particularly smart, but simply that science and math can indeed be mastered by ordinary people with regular backgrounds when they have good teachers and an environment that helps them value intellectual life. Those general conditions are present at essentially all of our nation's colleges and universities, a fact that means young Americans can learn as much science as they choose. And the corollary of that idea is that American citizens at all points in their lives can better understand the fruits of scientific research.

At this point it's worth acknowledging that some scientists seem happy enough to have their work appear to be so difficult it must be opaque to anyone lacking Superman's X-ray vision. That, in my opinion, is detrimental to our common lives together as citizens. It's a fact of the world today that the majority of research science is funded, in one way or another, by public money, clear reason that it is really in all our interests to understand the basics of technical research.

There's another gulf that some Americans believe lies between themselves and science as they perceive it. It's a simple but unfortunate feature of our social history that a few fundamentalist Christians took an early stance against both Darwin's theory of evolution and an ancient date for the age of the Earth. Most devout Christians, both here and abroad, never thought the book of Genesis required them to reject basic features of Earth's fossil record, of similar patterns in anatomy, of modern genetics, and more. But as a happenstance of American history, a small minority of Christians staked out a highly public position denying that religious life and modern science could both be embraced. That group has been masterful at generating a lot of noise in the public square in favor of its beliefs. As a result, many citizens who haven't thought deeply about science believe that it must conflict with their religious commitments. When I used to teach freshman geology I ran head-on into the ignorance not just of science in this respect, but also of common religious teachings. I had students from my classes earnestly tell me they were good Catholics, Lutherans, or Methodists and therefore couldn't accept the geologic evidence for the ancient age of the Earth or the fossil evidence for the transformation of species over time. As I was glad to explain to them, their denominations had, in fact, long since come to embrace both an ancient Earth and biological evolution.

But the misunderstandings between science and religion in our society cut both ways. Scientists, often betraying a remarkable degree of ignorance about simple facts, sometimes denounce religious life and the people involved in it in rather categorical terms. From my own experience as a church member involved with technical fields on a university campus, I can testify that some scientists are alarmed by religion, even when they may know nothing about it.

From my point of view, the misunderstandings between science and religion have seldom imposed as great a cost on society as they do today. Obviously it's a loss when some young people won't study fundamental scientific ideas because they think the technical concepts involved are evil. And when some scientists have roughly similar feelings about all religious life, gears that could easily mesh for the benefit of us all start to move in ways that clash. That's why some of my outreach work is meant to reassure my fellow church members that biology and geology don't constitute a threat to our ways of living. But equally importantly, I believe scientists—who are funded to a large extent by public dollars—cannot afford to needlessly alienate the people who write their paychecks.

For all those reasons I think it's important to try to translate technical research into a format all citizens of every age and background can understand.

My efforts at scientific outreach have been broad. I've led programs in public schools, at civic association lunches, during Cub Scout meetings, and in summer-time gatherings at public parks. In all those venues I've presented material about rocks, minerals, fossils, and the history of the Ice Age.

Then there are the nationally syndicated Rock Doc newspaper columns that are collected in this book. I started to write the columns when I worked fulltime for my local daily newspaper. I conceived of the Rock Doc and volunteered to write the column. I kept on writing the columns when I returned to university teaching. Later, when I took another university job, a wise dean gave me an assistant. Her diligent labors at promoting the column to other newspapers gradually built the Rock Doc circulation to over 200 newspapers from coast to coast. Throughout this time of enormous circulation growth, the column has been distributed for free, as a service of Washington State University

One thing is clear. I know I've been quite lucky in finding a way to marry my compulsive tendencies with a way of interacting with both my university colleagues and the public. Most days over the past 15 years I've rolled out of bed in the morning happy enough for a chance to lecture Rotary members, college freshmen, or the coast-to-coast reading public. Because all writers crave readers, the Rock Doc effort has surely been rewarding to me, offering hundreds of thousands of them each time I send out the columns.

While I'm grateful to have been as lucky as I have been, at a gut level I generally feel I'm flying blind when it comes to living. I've never had a clear plan of how to proceed down the dusty road of life, nor even how to make a living. Nevertheless, I've reached middle age without any sense that my professional life, with its self-appointed mission of science outreach, is any kind of yoke on my arthritic shoulders. By my lights, that makes me more than lucky and I try to be appropriately grateful.

But one potentially dangerous fact about being a compulsive writer is that there's a true paper trail of what I've thought, what I've tried to learn, and simply what's passed through my crazy head over the years. You, the reader, can judge as you look through this volume if I've spent my time to good effect. If I'm lucky, you'll find something of interest in these pages, and perhaps have a richer appreciation of science because I've labored at the keyboard again and again. In any event, I do appreciate your willingness to come with me through some of the compulsive internal journeys that lie behind these pieces.

SECTION 1
GEOLOGY AND PALEONTOLOGY

I may be a tad biased, but I think of geology as the queen of the sciences. The study of rocks, minerals, and fossils has led us to understand the rich history of planet Earth and the interrelated history of life.

The reason there is oxygen in the air for us to breathe is explained by paleontology. The ongoing risks of volcanic eruptions and earthquakes are also part of the geologic realm. Where to find gold and diamonds is another aspect of geology. In short, the field runs from the academic to the highly practical—and it never runs out of intriguing topics to explore.

Besides all that, it's a simple fact that 4-year-olds everywhere love dinosaurs and woolly mammoths. What could be better than studying what delights small children?

Geology and Paleontology

Stumbling on Good Finds
in Odd Places

About a decade ago I was cruising up a 130-mile-long reservoir behind Grand Coulee Dam, an enormous concrete edifice that sits across the mighty Columbia River in my native Washington State. The reservoir behind the dam is named Lake Roosevelt, in honor of FDR, who was president when the dam was built. The area around the reservoir is rural, but it's hardly the isolated wilderness of the Yukon. Nevertheless, geologists are still finding some quite intriguing things that lie in such rural places, simply because there are still many outcrops on Earth that we just haven't looked at seriously and from all points of view.

What had drawn me to the reservoir was evidence of Ice Age catastrophic flooding that raced across the Northwest from Montana, forming the landscape. The evidence for that story is abundant, with legions of outcrops telling the tale. Typically, we geologists drive to those outcrops in pickup trucks—it's part of our culture to drive everywhere in trucks (the larger, the better, we tend to think).

But on that particular summer's afternoon, I had left the world of pickups behind. The shores of Lake Roosevelt—for many miles—are steep bluffs. There are no roads above or below the bluffs. There aren't even trails that you could hike from the top of the bluffs to the bottom along the lake shore.

So I was in a motorboat.

I had borrowed the boat from family friends—only great kindness, I suspect, allowed them to lend it to a novice boater like me. But I managed to tow the craft to the reservoir, get it in the water, and start the motor. And on that fine afternoon, I cruised into an inlet and saw a big waveform embedded in the sands of the solid bluff. It was a stunning moment because my location that day, above the confluence of two rivers, told me I was looking at evidence of a catastrophic flood coming down from Canada to the north—not from Montana to the east. Voilà! I had discovered a reason to think we had catastrophic Ice Age floods coming from more than one direction in northeastern Washington State.

One of the most deeply gratifying things about doing fieldwork in geology is that such days occur—magical times when you find something quite new and significant.

More than a hundred years ago a much bigger geological find was made by a prominent American paleontologist just north of my own dear Washington State. A geologist named Charles Walcott was at work in the mountains of British Columbia looking for fossils from the specific time when animals in the sea first became large and complex. Earlier in Earth history, life in the seas had been very simple—not much more than worms and elementary jellyfish-like animals—but all that changed in what geologists call the Cambrian Period. Near the start of the Cambrian, rather suddenly by geologic standards, sea creatures appear in the fossil record that were much larger and had eyes and hard shells.

Walcott had spent the summer of 1909 in British Columbia, hiking to look for fossils in shale of early Cambrian age. One late summer's day he hiked up to a particular high-elevation ridge—and there at his feet were pieces of shale with a cornucopia of fossils that we geologists are still studying.

Walcott's rocks are part of what is called the Burgess Shale, and they've taught us how varied early life was in the Cambrian. Some Burgess animals had five eyes; others had complex pincers attached to their faces on their snouts. Some of the animals were so odd they seem like hallucinations—one has even been named "Hallucigenia" in honor of that idea. But, of course, each of the strange fossils were once real, flesh-and-blood animals in the Cambrian seas.

We don't know why the five-eyed animals went extinct and animals more similar to our basic body plan survived. Likely, there was a good factor of chance involved, and it could have come out quite differently. But one thing is certain—it was the two-eyed creatures that lived and became the animals that appear in all subsequent geologic periods.

Walcott's hike more than 100 years ago helped to propel the science of paleontology forward. It has taught us that animal life on Earth likely could have played out in many different directions. That's the fundamental philosophical issue worth pondering about Walcott's famous discovery. The history of life on Earth looks contingent on small, chance occurrences—with the odds against our existence stacked very high indeed. Perhaps that's a grand reason to rejoice. We are here, and able to study life forms of the past, having "won the lottery" that made two-eyed animals the norm.

Whenever you walk in the woods or motor along a lake's edge in a boat, keep your eyes open for gems of all kinds. The simple fact is that you may be the first person at a particular spot to seriously look for what will be literally lying at your feet. And from what I know of living, the hunt for something grand can be as rewarding as the finding of it.

Gold in Them Thar Hot Springs

Twenty-five years ago I was spending my summers beside sulfur-belching hot springs in northern California. To be sure, the hot springs were not as big as Yellowstone's. Most were just a few feet across, one or two about a dozen feet wide. None of them were truly boiling, but they were hot to the touch and smelly gases bubbled vigorously out of them.

To add to the general ambiance of roasting sulfur, air temperatures in that part of California each July and August are in the 100-degree range, and in addition to gases, the hot springs carry a lot of mercury, arsenic, and other toxic metals in the water. In short, any normal person would have fled the scene.

But I loved it all because the hot springs carried trace amounts of gold. That meant they precipitated meaningful amounts of gold where the waters cooled and the gases dissipated.

That's pretty unusual. In fact, that spot in California is the only place I know of where gold makes it all the way to the surface of the Earth in such low temperature waters.

Memories and images of those summers crowded my mind recently as I was learning about something else that's unique to certain hot springs, namely those that are more acidic and have a higher temperature. Yellowstone has some like these, with pH values down to 1 and even 0. If you have been to Yellowstone, these are the "ugly" and stinky hot springs and mud pots that lie to the north of the main lake.

It's natural to think nothing could live in the conditions of such spring water. But the Yellowstone hot springs have a host of "archaeal" organisms living in them. That name means they are single cell creatures that—like bacteria—have no nucleus, but they are even more primitive than bacteria.

One kind of archaeal organism is *Sulfolobus solfataricus*. (Yes, I know microbiologists like to assign challenging names to life forms they study. But just concentrate on the fact that the 'sulf' and 'solf' syllables suggest a love of sulfur, and the 'lob' syllable suggests a lobe-like shape for the microbe when you look at it under the microscope.) *Sulfolobus solfataricus* live in Yellowstone's hot springs and elsewhere in the world, too.

Professor Cynthia Haseltine of Washington State University and her graduate students study the creatures in question. She was kind enough to teach me a little about them because of my old interest in hot springs.

In middle age, I thought I knew the basics of life. But it turns out I'm woefully ignorant. Haseltine's microscopic "bugs" can fuel themselves in two radically different ways. Like you and me, they can eat carbon and other nutrients and live on that. But they can also make a living off carbon dioxide and sulfur. I knew little bugs could do the carbon dioxide and sulfur diet, but I didn't know they could switch back and forth to "our" sort of eating, too, when it suits them.

"But that makes sense in evolutionary terms," Haseltine says. "If a microbe is living in the hot spring with sulfur and carbon dioxide available, it can grow—but that's slow growth. But then if leaves fall into the hot spring in the fall, *Sulfolobus solfataricus* can 'switch' and make a living and grow more quickly using the leaves as fuel."

Now let's go back to California long ago, crouching over the hot springs that carried gold. (Kneeling and crouching was possible during that long ago golden time when my knees were good.) In a nutshell, all the work I did as a graduate student tried to explain gold transport without respect to living creatures, with reference to what is called inorganic chemistry.

But what excited both Professor Haseltine and this rock head while talking to each other is the possibility that some *Sulfolobus* organism is not just present in the spring—Haseltine is sure it is—but that it may be interacting with the trace amounts of gold in the waters in ways that bring them to the surface.

Imagine that: life forms that can switch from eating sulfur to dead leaves to, just maybe, ingesting gold, too.

That's quite a bug.

Blown Away by a Handheld Device

I parked my ample butt on the granite steps and waited in the shade of a campus building. As good as his word, Dan Hanson of Olympus Innov-X came to show me a real-life device that reminded me of Spock's tricorder on *Star Trek*.

When I was a young geology student many moons ago, I used to study the variation of chemical composition in the Earth's many rocks, minerals, and soils. Different elements could spell the route to finding a new platinum deposit or tracing out plumes of pollution around old chemical plants. In my day the effort required to identify chemicals meant days or weeks worth of work, taking samples back to the lab and painstakingly analyzing them.

The world has changed.

X-ray florescence, or XRF, is one way of determining what elements are present in a sample of material. The astounding thing to me is that the old XRF machines (weighing a ton) can now be replaced for many purposes by a machine not much larger than a workman's lunchbox. The "tube" source of X-rays in the device is similar to what powers medical devices, but on a much smaller scale.

Dan had his device in a small, black suitcase. He described how the instrument had been miniaturized since my days in the lab. Then he held the device against the granite on which we were sitting and pulled the trigger. In less than a minute, it provided data on the chemical elements present in the rock. Knowing granite as I do—it's a common rock—I wasn't surprised at the readout. But I was more than impressed with the speed of the analysis, complete with the minor elements the device was identifying.

It seemed to me that I was looking at Spock's tricorder made real in the present day.

Dan and I next stepped over to a very different kind of rock that my university had conveniently used to make nearby benches. Again, we held the device against the stone and pulled the trigger. The data rolled in.

Finally we wandered over to sandstone around the entryway of a building. We analyzed that, too. Total time elapsed for three pieces of impressive work: about five minutes. Clearly Dan's device could save a company or a prospector oodles of money on assays compared to the old way of doing business.

The handheld device has some limits. It's much better at certain applications than others. It does very well, for example, identifying the metals in alloys.

"The scrap iron business is one of our better customers," Dan told me. "They use it to identify stainless steel on the spot, so it can be sorted into a different bin from the rest of what they process because it's worth more. For that type of work, the analysis takes only half a second."

Another example of what the tricorder does well is helping with major demolition and salvage projects.

"One of our customers paid for his on the first day he had it on the job. They were taking apart an old chemical plant. The solder that had been used on the pipes contained silver. So they could pull that out separately and sell it," he said.

The basic model of the XRF costs about $22,000. Part of that cost is the X-ray "guts" of the tool, and part of it pays for the complex software that calibrates the device and can be changed for different applications. Another top of the line model provides an analysis called X-ray diffraction as well. That means the tool can tell you not just chemical composition, but which minerals are present in the sample. To me, that's like one miracle compounded by another.

Although Dan works with the device every day, he's still impressed by it.

"We don't know yet all we can do with it. There's more to learn," he said with a smile.

I'm always impressed with the fact that science and technology continue to make progress even though other parts of our collective lives all too often don't. To put it another way, we don't get better at writing novels, composing symphonies, or forging constitutions—but science and technology take major strides forward each year. The success of technical research is one of the reasons I'm still quite hopeful about the world's future, despite our current economic challenges.

How Fragile is the Solid Earth
Beneath Our Feet

Geology is often in the news, with the price of petroleum sometimes fluctuating up or down.

But matters became suddenly worse in just a few minutes when Earth processes sprang into the headlines in full destructive force.

Even geologists like myself, used to the ferociously destructive power of earthquakes, were taken aback by the tragic news from Japan on March 11, 2011. The largest seismic event since earthquakes were first measured in that nation, near a 9 on the Richter scale, devastated sections of the northeast coast, and major aftershocks rocked the region for days thereafter.

The epicenter of the massive quake was under the sea, and a tsunami was immediately triggered by the event. The word "tsunami" has replaced what older readers may remember as a tidal wave, a name that was highly misleading because tsunami are not caused by the tides. The name tsunami is Japanese, a fact that shows Japan has been plagued by earthquakes and tsunami for as long as Japanese civilization has existed.

Tsunami are usually caused by movement of the solid sea floor, a lurch either up or down, that sets an enormous body of water on the move. The water packs a great deal of energy, like an enormous sledgehammer.

As a tsunami flows into more shallow conditions near the coast, the height of the wave increases more and more. That's why ships far out to sea are not tossed by massive waves, but people in a harbor can see a truly enormous surge of water coming toward them. The water can spill far inland, as it did in northeast Japan.

Tsunami travel fast—at literally hundreds of miles an hour. Because of that fact, there was little time between the quake itself and the tsunami hitting the coast of Japan. Much of the destruction of the quake was from the effects of seawater inundating the land, sweeping whole buildings off their foundations, undermining roads, and, most unfortunately of all, quickly sweeping thousands of local residents to their deaths.

Because tsunami travel across the entire Pacific Ocean, damaging coasts thousands of miles away from the original earthquake, scientists have a tsunami warning system in place 24-7. Nothing is perfect, but it's a good system, and warnings in Hawaii preceded the arrival of the tsunami there.

As usual with massive earthquake damage, fires broke out in Japan and burned out of control in some cities. Fires often follow major seismic events because natural gas pipelines crack and start to leak, and because electric lines fall and create sparks. To make matters worse, firefighters can be hampered in their work because water mains are broken.

We must all wish the Japanese people well as they struggle to cope with what has happened. That's a process that will play out over years to come and give Americans an ongoing opportunity to help our neighbors across the Pacific.

But in the United States, we should also take time to reflect on the fact that two or three large regions of the Lower 48 stand at risk of similar events—and we are generally less prepared than the Japanese to deal with major quakes.

California most famously, plus Oregon, Washington, and inland western states like Nevada, are all slated for massive quakes in the future. But it's also true that the central part of the country, where Missouri, Kentucky, and Tennessee come together, is another place where geologists are sure there will be massive quakes. Even South Carolina is a hot spot for seismic risk.

We've got to learn from what happened in Japan. We can do better when it comes to everything from protecting our infrastructure through government regulation to having personal family plans in place for emergencies.

Let's use the tragic event in northeast Japan as a wake up call right here at home.

Norsemen and the Great Crack in the Atlantic

More than a thousand years ago, my Norse ancestors were busy pillaging Europe. Using Viking long boats (the stealth weapon of the day), we could sneak up the coasts and rivers of civilized areas, plundering and pillaging at will.

But perhaps we were not thoroughly evil souls. After just a couple of bloody centuries, we converted to Christianity, beat our swords into plowshares, and became peasant farmers. From that day until this one, we've been known only for eating fish prepared with caustic lye and making stupendously depressing films.

A few Norse arrived in Iceland about 1,200 years ago—quite far back, indeed. They realized they needed a way of settling disputes on the island, so they began holding open-air assembles at a spot called Thingvellir, a location where some dramatic, uneven, and wide "cracks" in the rocks provided natural highs and lows that were useful as speaking platforms and seating areas.

In a very rough and ready way, the Icelanders had started grassroots self-government. Instead of physically attacking your neighbor, you could come together with others and seek a resolution to questions like which sheep belonged to whom or which gods should be worshipped in Iceland.

Ancient Iceland was far from a perfect or sophisticated democracy, to be sure, but it can be seen as a step forward from anything northern Europe had seen earlier in history. That's why the Thingvellir location is a historic shrine in Iceland today.

Every geologist who visits Iceland goes to Thingvellir. It's not that we are good students of democratic history—it's because the great cracks of Thingvellir sit astride some major volcanic features.

Thingvellir is located on what geologists call the Mid-Atlantic Ridge, a rift tens of thousands of miles long that runs the long length of the Atlantic and vomits up enormous volumes of lava. The ridge is mostly underwater, but in Iceland it has risen above the waves (convenient for the Norse, so they could settle it).

The Mid-Atlantic Ridge is a place where two enormous tectonic plates move apart from each other. They creep along at about the rate your fingernails grow. That doesn't sound like much, but when you consider that a whole region of Earth is moving, you can get a sense of the great magnitude of the forces involved in the process. The movement of the two plates away from each other allows molten rock from deeper in Earth to well up toward the surface where it becomes lava. When lava cools, volcanic rock is born.

At Thingvellir, the ancient Norse actually used landscape features—natural platforms on which to stand, long and low places on which to sit—created by the Mid-Atlantic Ridge to help make their rough-and-ready democracy work.

But there's also now another connection between Icelandic society and geology. The tiny island nation is energy-rich because it has abundant heat stored in the young volcanic rocks under the island.

Many Icelanders heat their homes with water that is heated in the rocks beneath their feet. And Iceland generates electricity from turbines turned by steam from the same source. There are risks to living so close to Mother Nature's warm heart—namely volcanic eruptions—but real benefits, too.

Iceland got clobbered by the financial "bubble" that swept so much of the world in recent years. For reasons I don't quite fathom, institutions and individuals in Europe sent their money to banks in Iceland offering high rates of returns. But those banks are long since belly up and the country itself became bankrupt. (No, I don't understand how a country can be bankrupt. A mere geologist has her limits.)

As an old Viking myself, I've got to be hopeful that Icelanders will soon find ways to better use their amazing reserves of heat energy from volcanic rocks to work their way back to solvency. One hope is to generate geothermal electricity and ship it to Europe via underwater cables.

With one of the oldest democratic heritages in the world, and surrounded by stunning geology, Icelanders have a lot going for them. Let's hope it's enough to get through these tough times.

Chance Discoveries in
Our Backyards

Some of us have never found a dollar bill on the sidewalk. But once in a blue moon a lucky soul in the upper Midwest reaches down into the Ice Age deposits of the region and plucks out a diamond.

It doesn't take much geologic knowledge to recognize a diamond in the rough as an interesting and valuable object. Diamonds are the hardest mineral in the Earth, which means they will scratch quartz, window glass, and other hard gems like ruby and emerald. And diamonds have a very high luster, a term referring to their ability to reflect light.

In short, it's not too tough to identify diamonds, even before they have been shaped into cut gemstones.

That's why, since the 1800s, farmers and other residents of the Midwest have occasionally spotted and scooped up a diamond when digging water wells or otherwise disturbing the ground.

Many diamonds around the world are found near places where deep Earth processes blasted them to the surface in special rock called kimberlite (named for Kimberley, South Africa). But there are no rocks like that in the Midwest.

Instead, the diamonds in the central part of our country were transported to their resting places by the enormous glaciers that dominated North America during the Ice Age. Geologists spent several generations looking for the ultimate source of the diamonds "up ice" in Canada. It's only relatively recently that a few dedicated—not to say obsessed—geologists found the sources in northern Canada.

Geologists usually look for diamonds not by initially searching for the gems themselves, but by looking for more common minerals that often come along with diamonds from their deepest sources in the Earth. Even so, finding diamond-rich rocks is generally a needle-in-a-haystack challenge. But, in time, dedicated exploration geologists found the source in northern Canada of the diamonds that are—occasionally—found in our upper Midwest where the ancient glacial ice left them. The fascinating tale of the search is one you can read about in the book *Barren Lands* by Kevin Krajick.

Other people discover objects of no monetary value but that mean a lot to those of us interested in Earth history. That was the case earlier this summer when a contractor in Tennessee digging eight feet down for a swimming pool unearthed a fossil. It appears to be the jawbone of a trilophodon, an extinct relative of the mastodon.

Mastodons and their kin roamed the Earth during the Ice Age—the same time that a few diamonds were arriving in the Midwest courtesy of Canadian sources. You'll probably remember mastodons (dimly) from childhood books or posters about the Ice Age. Mastodons were browsers rather than grazers, a point scientists can deduce from the shape of their teeth. The woolly mammoths, in contrast, were the large and famous grazers of the era.

All of the mastodon-related species went extinct as the Ice Age came to a close. That may be because enormous, natural climate change was sweeping the Earth. And it may be that many animals in North America were also facing increased pressures from human hunting.

No one is well-prepared to find Ice Age fossils when digging for a swimming pool. At first the folks in Tennessee thought they had found the jawbone of a dinosaur. Homeowner Jim Leyden got a call from his wife reporting exactly that. But a conservator from the local Pink Palace Family of Museums steered the understanding of the discovery toward the mammals of the Ice Age. (The Pink Palace features a wide variety of exhibits ranging from natural science discoveries to a replica of "the first self-service grocery store in the country, Clarence Saunder's Piggly Wiggly.") The Pink Palace conservator estimated the fossil trilophodon weighed up to two tons as a flesh-and-blood animal.

The discovery in the swimming pool pit was quite a surprise to all concerned. According to a report from the Memphis *Commercial Appeal*, Leyden said, "I grew up in New Jersey. I might find a body, but not a prehistoric animal."

Leyden told reporters he planned to donate the jawbone to the Pink Palace.

As he said to Memphis WMC-TV, "What am I going to do with it? If I keep it around, my wife might throw it at me."

Seizing the Sunlight

N othing is sweeter than a powerfully sunny Saturday that comes along in autumn. And some of us rock-heads think that perhaps the very finest way to spend such a fall Saturday, one stolen from the relentless turn of the seasons, is looking at gravel that reveals Ice Age history.

On a recent Saturday, several friends and I loaded my car with cold chicken, hot coffee, and a few topographic maps. The object of our quest was the evidence of violent change authored by Mother Nature a few thousand years ago. I've made the particular trip we had planned many times—far too many to count. But each time I go, I see something new, and I'm genuinely glad to show others the tangible geologic evidence of massive, natural climate change.

There was a day when geologists were confident that every kind of natural change in geologic history was always gradual. Rivers and glaciers, we figured, must always move as slowly and predictably as we see them around us today. The shape of the land's surface—the ups and downs around you wherever you live in the United States—could be explained mostly by the gradual erosions of streams. The slow erosion of glaciers completed the picture in the northern tier states and at high elevations in the Rockies.

It was a perfectly good way of looking at the world, at least for a long time. But the evidence of the land and gravels in the central portions of the state of Washington don't fit into that picture.

Essentially, I happen to live next to one of the most interesting places on the planet for studying how unkind the effects of climate change were 14,000 years ago. In my neck of the woods, every gravel pit or river bar can show the interested observer the effects of cataclysmic change, forces that revolutionized the landscape on the scale of hundreds of miles over just a few days.

The Mother-of-All-Landscape-Change, if you will, stemmed from an enormous glacial lake that existed in the late Ice Age in what is now Montana. Called Glacial Lake Missoula, the lake held about as much water as Lake Superior. But unlike Lake Superior, Lake Missoula was at high elevation—in the Rockies—and it was held in place not by the Earth but by a dam in the Clark Fork Valley of Idaho that was made of glacial ice.

Obviously, ice dams are a tad dicey. Glacial ice always advances and retreats, and ice is more fragile than rock. The day came when the ice dam melted or broke—as it simply had to at some point—and all the waters of Lake Missoula in Montana started to move downstream and toward the west.

Lake Missoula created no ordinary springtime flood, but inundated whole sections of the Northwest. At first, the water depth was measured in *thousands* of feet—because such was the depth of the former lake. The torrent carried boulders the size of houses, gravel, sand, and rafts of ice, all in the waters and moving downhill at the speed of a freight train.

When the flood waters hit my native Washington and started to spread out, they were still hundreds of feet deep. They stripped away our soil and carved the land down into the bedrock creating "coulees," or flat-bottomed gashes in the Earth. To this day, the coulees crisscross what we call the Channeled Scablands, a landscape that only a native can love (but that I surely do!).

The floodwaters continued on their journey, hitting the valleys of the Snake and Columbia Rivers. The tremendous erosional power of the mega-flood helped to enlarge the Columbia River Gorge that cuts through the Cascade Mountains. Finally, the floodwaters were discharged into the sea.

Mother Nature wasn't sweet and gentle in the late Ice Age. If you want to learn more about the tangible evidence for her fiercest moods yourself, you can spend some winter evenings reading by the fire about the events just sketched. (There are both books and zillions of internet offerings on the topic, too many to list here.) Then, next fall, come to the Inland Northwest and see for yourself.

With glorious sunny days and cold chicken lunches thrown into the mix, what's not to like?

From the South Pole to the Northern Arctic

Geologists often travel to remote locations. One of my geological friends has lived in a yurt in Mongolia while exploring for metals, and another works scouting for ore high in the Andes of Chile. But my friend Mike Pope takes the prize for traveling to the back-of-beyond. He has slept in double-walled tents at high elevations in Antarctica, and he has hiked up slopes of the McKenzie Mountains in northern Canada.

Pope's travel tales are second to none, and his journeys have helped him piece together the story of life itself over immense periods of geologic time.

Pope studies fossil evidence of the most dramatic set of changes in life on our planet. Most of Earth's long history happened before any animals lived on land. In fact, 90 percent of the total span of Earth's enormous history was in this period, with only simple life forms like bacteria and worms in existence, all living in the seas.

I'd like to be able to tell you how scientists explain that life remained so simple for so long, then rocketed into complexity in the last 10 percent of the story. But I can't. We just don't know why history transpired that way. Still, the basic facts themselves are clear from the fossil record, with important details in the story being added by new discoveries in remote locations like those that Pope has explored.

Scientists like to lord it over other people sometimes, but the truth is it took us geologists a heck of a long time to see that there were fossils in the rocks deposited in the long but ancient part of Earth history. For many generations, we literally stepped over the little markings that the worms left behind.

To give you a sense of what the ancient impressions from the first 90 percent of Earth history are like, let me mention that I can see the marks when I have my bifocals on my nose, but not otherwise. Still, once we learned to see what was there, geologists rather easily started finding the fossils in Earth's oldest rocks all over the planet. (Just an example of the basic fact that humans see best when they know what they're looking for.)

Pope has studied the great shift of animal life on Earth as the little worms that left the faint impressions were replaced with abundant animals that had shells, teeth, and bones. From that point onward, fossils are objects I can see without my bifocals, things obvious enough that it's child's play to recognize what is underfoot.

Once the big and obvious fossils appear in the rocks, the plot in the story of life accelerates enormously. Giant amphibians are soon followed by reptiles, then dinosaurs and birds—and the wild successes of our own group, the mammals. That's the part of life's history that school children delight in, from enormous flying reptiles to dinosaurs and wooly mammoths.

Pope reminded me once that, long ago, when reptiles were the new kid on the block, an interesting series of changes led to severe climate change. There was a lot of carbon dioxide in the air back in the day, much more than anything we will see in our atmosphere in the centuries to come.

Pope has been studying rocks created during this time of fluctuating carbon dioxide. The evidence shows that glaciers advanced and retreated in many dramatic pulses as carbon dioxide levels dropped from their earlier high levels. The changes were the opposite of stately, with shifts back and forth as common as dirt. That's a pretty good parallel to what we will see in the coming centuries and millennia, Pope thinks, except that we are now headed in the opposite direction due to burning fossil fuels.

The Earth plainly teaches us that the only true forecast about climate is one of constant change and dramatic reversals. There is no comfort in that lesson, but people like Pope have gone to the ends of the globe to establish what we can know of the future by looking at Earth's past.

The Downside of Good Fortune

The rate of China's economic growth is often reported in the news. But China is experiencing another bonanza, too. It doesn't get the headlines commanded by economic figures, but it catches the attention of geologists and anyone with an interest in the history of life on Earth. What's at issue is the absolute tsunami of fossil specimens that are being dug up in China and making their way around the world.

The main period for discovering important fossils in the United States is probably in the past. To be sure, there are dinosaur finds in Montana from time to time, and other fossils come to light on occasion around the nation. But simply because we've had scientific expeditions looking for fossils within our national boundaries for well over a century, the rate at which we discover significant fossils these days is rather low.

In contrast to the United States, China didn't have people systematically looking for fossils within its borders until quite recently. Locked in poverty and separated for a time from much of the world through the Cold War, the Chinese had other issues to deal with besides trying to unravel the ancient history of life on the planet.

But all of that has changed in recent years. Both scientific and private efforts have been mounted to find fossils in certain Chinese rocks, and the efforts have been enormously successful. The flood of fossils that has emerged is flowing into Chinese museums, into other museums around the world, and into the hands of private collectors.

There is a lot of money at stake these days in the world of fossils. A complete specimen of a transitional type of dino-bird species fetches a king's ransom on the international market. Ditto for certain specimens of enormous swimming reptiles or pretty nearly any other species that is large, interesting, or fierce-looking.

But here's a simple fact about fossils. Usually what you find in a rock is only part of one individual's remains. In other words, it's not so hard to find a part of a dinosaur—I stumbled across one isolated dino bone myself in Montana when I was younger. But it's truly rare to find all the bones and teeth of a particular individual, laid out neatly and waiting for you.

There often is real scientific value in just part of a fossil. Incomplete specimens can mean a lot to experts. But, for obvious reasons, the price of a complete fossil is extremely high compared to the price an isolated bone or tooth will fetch.

Many of the fossils being found today in China first come to the attention of farmers who simply find a rock in their field with part of a fossil in it. From the point of view of some farmers, adding to that first rock with part of another is not so difficult. For example, imagine you and I both become fossils in the same siltstone bed, but the fossil version of me ends up missing an arm because a predator carried it off after my death. It might occur to a farmer to "add" an arm to my fossil remains from another partial specimen—like the chunks of you that were fossilized. This creates a mixed fossil, one that represents the same species, but that has two individuals combined in it.

Even more problematic is that a farmer might not have another human bone to complete the fossil version of me. In that case, he might add whatever he has on hand that seems most suitable, something from another species of animal. This creates a chimera, an animal that never lived at all.

Farmers and local fossil dealers can fetch high prices for specimens they create in such a way. And according to a recent article in the journal *Science*, it looks like the majority of fossil specimens coming out of China have been doctored.

It's all a crying shame. It would be better to have incomplete evidence of an animal than to have "specimens" that have been carved, forced together, or otherwise corrupted.

I surely don't begrudge Chinese peasants any extra income they can get. But I do regret the loss of scientifically meaningful evidence of life's rich history on Earth.

More Shaking from the Earth

Many Midwesterners don't know they live in a region where earthquakes can strike. But some residents get small reminders of that simple fact from time to time when quakes near 4 or 4.5 on the Richter scale rattle a few nerves. Every time such a mini-trembler is reported, I like to hope we will all learn from it—because the center of our country is woefully underprepared for what will come in terms of later, much larger quakes.

As students of American history know, the previous "Big One" that struck in the Midwest happened a good while back, in the winter of 1811–1812. The quake was really several enormous events that occurred in and around what is now Missouri over three months. Some of these mega-quakes were felt as far away as the East Coast, where church bells rang from the shaking. Parts of the Mississippi ran backward for a time due to the upheavals. The shape of the land itself was changed for many square miles, some of it rising upward, some sinking downward to become swamps. The big quakes were accompanied by hundreds of small ones, and local observers said that at one point the ground shook almost continuously for weeks.

The monstrous events are known as the New Madrid earthquakes—the name comes from a small town leveled in one of them. Modern geologists would say their epicenters were all part of the New Madrid Seismic Zone, which is an area that falls in the neighborhood of where Kentucky, Tennessee, Missouri, and Arkansas all come together on the map. Because modern seismographs had not been invented by 1811, there's no way to say exactly how big the quakes were. But based on descriptions of the effects of the quakes, geologists estimate that they were in the 7 to 8 range on the Richter scale.

What do these numbers mean? You can expect areas near the epicenter of a Richter 8 quake to generally experience what is termed "total" damage. Because of the way the Richter scale works, a value of Richter 7 corresponds to a release of 31 times less energy than a Richter 8 quake, and is expected to result in mere "major" damage. And as another point of reference, of all the last century's most deadly natural disasters the world around, about half were earthquakes—because even Richter 7s are deadly in the extreme if population is dense and buildings are made of bricks and concrete blocks.

One of the most staggering things about the New Madrid winter of 1811–1812—sobering even to us geologists—is that there were *multiple* enormous quakes. Each one of the events would be a huge disaster today, affecting major cities, transportation and communication systems, small towns, rivers, and levees. A full repeat of all that happened in New Madrid would be a greater disaster than anything we've ever seen out here along the west coast, where this Rock Doc hangs her hat. Certainly the great quake in San Francisco in 1906 would pale in comparison.

A confounding factor about the current New Madrid risk is that the risk increased so much in the 20th century as population in the area grew. That is one reason everybody in the Midwest—just like us westerners—should practice what they want their family members to do in the event of a major quake. And, in my opinion, stores of water and food are very much in order in the New Madrid Seismic Zone and the areas around it, just as they are along the Pacific Coast.

We live on the outer crust of the Earth, and it's a fragile place to be. But you can do basic things to help yourself—and they can be useful whether you face an earthquake or, perhaps, another kind of emergency. Google "Red Cross emergency kit," or make your own stores of basics, like this rock-head does.

Because another Big One is coming.

SECTION 2
ENERGY, ENGINES, AND AUTOMOBILES

The fundamental physical reason we live in a wealthy society is that we have at our command large amounts of energy in many forms. Our fossil fuels are petroleum, natural gas, and coal. We've learned to supplement fossil fuels with smaller amounts of hydropower, nuclear energy, wind and solar power, as well as some biofuels.

The American energy landscape is changing. Politics is interlaced with how we choose to shape what energy resources we use. But no matter our political commitments, basic physical science will govern what we can do to both harness energy and use it as efficiently as possible.

Energy, Engines, and Automobiles

One Law We Don't Ever Break

I was surprised by how terribly hot the engine smelled.

On a recent Saturday I drove my '87 pickup truck home with a "new" camper on its back. I bought the camper for $800, about right for one built in 1975. I'm pleased (and surprised) to report the propane fridge and stove still work, so I was in high spirits as I started up the highway from the Snake River to the plateau where I live—some 1,200 feet higher.

When I smelled the hot engine, I realized with a sigh that I had no water with me. (You can trust a Ph.D. not to bring water when it obviously would be wise to do so.) So I turned on the cab's heater to draw heat away from the engine, and pulled over repeatedly along the side of the highway to let the engine cool at idle. I made it to the top of the grade with a hot engine and hot legs, but no boiling over.

Pulling over along the highway gave me plenty of time to reflect on why we physical scientists have studied and loved engines ever since Europe's first great energy crisis.

Here's the story:

By the mid-1600s, the people of Great Britain were in serious trouble. They had, essentially, burned most of the wood they possessed in their cool and wet country. They were staying alive huddled over peat fires, but soggy peat is just about the world's worst fuel. Where could the Brits get the heat energy they desperately needed?

The answer lay under their feet. The British started to seriously mine coal, following the seams into hillsides and down into the Earth, with most of the work done by hand.

One problem with underground mining is that you have to get the "good stuff" up to the surface, which is a lot of work in itself. And you also have to pump water, 24-7, up and out of the mine so your miners don't drown before dinner. The British used both horse and human muscles to do all that hard work—until an ironmonger (no Ph.D. he!) had a much better idea.

Thomas Newcomen came to realize he could build a piston-driven machine that would use the heat from burning coal to do the most brutal labor of mining itself. His was the first modern steam engine, a special type of heat-to-work machine about the size of a small house. The power stroke came from atmospheric pressure, the kind of air pressure in which we all "swim" every day.

That engine, in time, was replaced by high-pressure steam engines you've seen in historic locomotives. These were heat-to-work machines so greatly decreased in size they could move around on a track, literally hauling the world into the modern era.

Pure science, which had been sadly lagging behind during parts of this story, caught up quickly in the 1800s. Work and heat—which had been thought of as quite separate entities—were soon understood to be forms of just one thing. The stuff that work and heat made up came to be named "energy," a single concept that had not existed earlier. This was revolutionary physical science at its best. Soon, electrical and chemical energy were added to the list of forms of energy that can be transformed, one to another, but not created or destroyed. That's the often quoted First Law of Thermodynamics, a formal way of describing energy science.

Hauling my camper up more than a thousand vertical feet of highway, my engine had to use a lot of heat from chemical energy (stored in the gasoline) to do a heck of a lot of work. And because, sadly enough, engines are only about 25 percent efficient, three-quarters of the gasoline's heat ended up roiling out of the radiator and engine block—and onto my bare legs.

The science of energy explains much of the natural world around us, from the biochemical reactions in your cells to what happens in high-tech electronics. But it was the study of desperately needed, if crudely constructed, heat-to-work engines that first led us to truly understand energy—the greatest and most flexible of all our resources.

Hats off to ironmongers!

Saving Gas Upside Down

If you've been feeling despair about your gasoline bills recently, you have me for company. Having a root canal is fun compared to filling up the dual gas tanks of my old pickup.

It's no wonder lots of us are examining our driving habits and our vehicles with new intensity. Here's an idea that can really help your family sharpen its pencil and make changes in driving that will really matter.

Over the years we've grown used to thinking about fuel efficiency in miles per gallon, because that's the way the EPA reports it. My sedan gets 27 miles per gallon, your compact car gets 31 miles per gallon, and so forth. That's valuable information. But if you own two or more cars, standing those numbers upside down could help you save gas.

If your family owns two vehicles that you drive about the same distance, and if you are considering turning one in on a more fuel-efficient model, then the most useful way of comparing vehicles is actually gallons per mile rather than miles per gallon. With the help of an example, let me show you how it works.

Let's say—just hypothetically speaking—that your family owns a Hummer that gets 10 miles per gallon and a compact car that gets 33.3 miles per gallon. And let's say you drive each vehicle 300 miles per week.

For the Hummer, you are buying 30 gallons of fuel per week (because 300 divided by 10 is 30 gallons). For your small car, you are buying 9 gallons of fuel per week (because 300 divided by 33.3 is 9 gallons).

Say that you'd like to turn in one of the vehicles for a more fuel-efficient model. Your teenagers want to turn in the Hummer for a smaller SUV that gets 20 miles per gallon and looks great. Your spouse wants to turn in your compact for a hybrid that gets 43.3 miles per gallon. In other words, both choices mean you will get 10 additional miles per gallon from one of your vehicles.

Which should you choose? The increase is 10 miles per gallon in each case, so you might think there's no difference. Should you flip a coin? On the other hand, the 43.3 mpg figure looks great, so maybe your gut says to go with the hybrid.

If you want to save gas, here's how it works out.

If you turn in the Hummer for the smaller SUV, you'll use 15 gallons each week to fuel the new vehicle (because 300 divided by 20 is 15 gallons). That's a savings of 15 gallons over where you were with the Hummer.

If you turn in your compact for the hybrid that gets 43.3 miles per gallon, you'll use 6.9 gallons of gas each week to fuel the hybrid (because 300 divided by 43.3 is 6.9 gallons). With your permission, I'll round 6.9 to 7 gallons, to keep the numbers simple. So, that's a savings of 2 gallons compared to the 9 gallons you were using for your plain-vanilla little car.

Clearly, you'll save a whale of a lot more gas by going with the SUV to replace the Hummer. The allure of the hybrid's fantastic EPA rating is sweet, to be sure, but don't let it confuse you about which is your best option under the circumstances.

I recently bought a vehicle to replace one of two that are in my personal fleet. Oddly, perhaps, I chose to retain the more inefficient beast I own, my 1987 pickup with a V-8 engine. But I did so because I drive the truck less than 50 miles per month, so its inefficiency isn't really a significant issue. I also happen to believe, as a matter of deep spiritual conviction, that a geologist should be separated from her truck only by the angel of death. But that's just me.

Here's the bottom line: If you drive two vehicles about the same amount each month, take the time to do the gallons per mile analysis of your situation following the example above. Your reward will be making better decisions when looking at replacements for what you're driving today.

By the way, if you want to disorient a young car salesman, do your gallons per mile calculations on some scrap paper right on the showroom floor. If you can do arithmetic the old way (by hand, imagine!), the disorienting effect is even greater.

No matter what, good luck at the gas pumps.

Powering Up the Next Generation

When I came into the world in the 1960s, some visionaries still hoped that electricity would become "too cheap to meter." If you've looked at your utility bill lately, you know we are far from being awash in any such socialized abundance.

But it's also true that solar energy—simple sunlight—is greater than any other energy source we might be able to harness. Each July, in particular, it's clear at a gut level how much total energy blasts its way down from the skies onto our roofs, our cars, and our sweaty little selves as we mow the lawn.

But can we find a way to economically capture and use the energy that reaches us each day as sunlight?

Think for a moment of the familiar solar panel bolted to the roof of an RV. I've got one on my beloved 1972 travel trailer. It's a complex system made of expensive pure silicon that's been "doped" with even more expensive and toxic trace metals. The panel generates electricity—but only at a relatively high cost—making it most useful only for small applications or when access to the power grid isn't possible.

At least that's been the case until recently. With the increasing cost of electricity from the grid and the decreasing price of solar panels in recent years, the world is changing.

It can now make financial sense for a homeowner to plunk down something like $20,000 to $25,000 for solar panels for the south-facing roof of a house. If you do that, your power meter will run as usual during the night, when you draw power from the grid, but it will run "backward" when the sun is up and your rooftop power plant is making juice.

If you can make about as much electricity as you use, you can have no net electric bills for the life of the panels, roughly 25 to 30 years. The panels can make good sense in financial terms if you live in your house long-term and if you expect the cost of electricity to rise and rise through the years—as I surely do.

But the news gets even better. Rather than continue to accept the high cost and toxic components of traditional solar panels, some research chemists are looking toward a next generation of panels. Some of these devices are based on common organic molecules from the plant kingdom.

The basic idea has the appeal of simplicity: who would know more about solar power than Earth's plants, which have been using sunlight for energy for about 4 billion years?

Energy, Engines, and Automobiles   29

Some current research efforts look for guidance and inspiration to the brightly colored molecules of botany—the greens and reds of plant matter found all around us. One step of the great transformation plants accomplish each day, as they convert sunlight into sugars, involves setting electrons loose—the basis for electricity. The moving electrons are associated with the green and red plant molecules. If you can tap that supply of electrons, you can draw off a tiny electrical current.

Obviously, we're still at square one in this work. But the good news is that even school kids and aging geologists—with appropriate coaching—can generate this type of electrical current using natural "dyes" from plants. And tiny currents are enough to make a real difference, if the ingredients are cheap and can thus be scaled up.

Professor Jeanne McHale is a chemist at Washington State University investigating several topics, including how the "dye" molecules that make beets so strongly red interact with light. McHale and her colleagues around the world are working on a variety of similar molecules to potentially produce what are called dye-sensitized solar cells. Keep your eyes open in business news for the phrase "dye-sensitized" and you can follow the progress of researchers in this arena in the years to come.

If next-generation panels can be perfected in one form or another, they would create electrical power without requiring the energy-intensive production of silicon and the mining and use of toxic metals.

Solar power could become both cheaper and greener in our lifetimes.

Much Better than the
Good Old Days

I like experiments that have dramatic results—preferably explosions or at least a good fire. As it happens, a series of excellent historical experiments helped give us energy resources beyond what earlier generations ever imagined possible. And I think the story of those explosions and fires predict the basic path of our next stride into an energy-rich future.

William Murdoch was an early Scottish engineer who had the same fondness for dramatic experiments that I do. He took the water out of his mother's teakettle, put coal inside it, and applied heat. Quickly enough he found that the gases coming out of the teakettle spout would burn. He had invented "coal gas" and laid the foundation for gaslight.

Coal gas is filthy, and it often contains carbon monoxide. But it was better than using whale-blubber oil for dim lamps—the common alternative at the time. Soon major cities had large industrial sites where coal was heated each day in great boilers. Pipes carried the resulting gases under streets for gaslight on the street and later in houses. Eventually some cook stoves in homes used coal gas for fuel.

From time to time coal gas created dramatic explosions and fires, and from time to time gas leaks containing carbon monoxide killed people more quietly—but such were the hazards our forebearers endured in exchange for light in long, dark winters and for a convenient way of cooking supper.

Running the lighting and cooking needs of whole cities on coal gas meant a lot of daily work to manufacture the fuel. Coal gas was a major industry the world around. But we geologists were doing some simple experiments and making deductions on our own, and in time we found a much cleaner flammable gas in the Earth. This gas didn't have to be made fresh each day in enormous boilers. Because the new gas wasn't manufactured, but was a natural product of the Earth, we called it "natural gas." Obviously, people the world around still use a lot of natural gas each day, for heating homes, cooking supper, and—more and more—turning steam turbines to generate electricity.

In the process of mining the Earth for natural gas, a smaller amount of propane is produced. You know propane as the bottled fuel that commonly runs backyard barbeques. My beloved 1972 travel trailer has a propane cook stove in it, but it also has a propane gaslight above the stove—a last "shout out," as the young people would say, to the old days of using burning manufactured gas for light. The lamp does work, but it immediately reminds me of the phrase "more heat than light" each time I fire it up.

I really only use the propane lamp for fun—because of yet another technological revolution.

A couple years back I bolted a solar panel to the roof of my aging trailer. It easily recharges an old truck battery, on which I run a high-efficiency fluorescent light that's more than bright enough to read by for hours.

A friend of mine who works for the National Park Service lives off the grid and without electricity in the woods of northeast Washington State. He and his wife raised two sons with a single solar panel that powered one light. That was the only illumination the family had for reading and doing homework each evening, all winter long. (By the way, both of the sons have now earned graduate degrees in engineering. They are hard at work contributing to the next round of technological advancements people will take for granted 30 years from now.)

Here's the bottom line. We are living better, step-by-step, because of the progress of science and engineering. Whale blubber, coal gas, and propane lamps are all in our past. They won't resurface because we have better options emerging from the innovation that's a daily habit at research universities and forward-looking companies across the country.

Hang in there. I know the economic situation is amazingly difficult. But I also know much better times are on the horizon.

Guaranteed to Make
the Lights Come On

On a windless night recently I arose before the sun. I flipped on the lights and plugged in my coffee pot, the old fashion kind that actually clears its throat to make a steaming pot of Joe.

While waiting for the coffee to percolate, I may have felt like an old rag and looked worse. But one thing is certain: I didn't spare a passing thought for the demands I was placing on the electrical grid.

In Detroit they speak of the Big Three automakers. But we have a much more essential Big Three in this country, namely the three sources of electrical power that always work, 24-7, no matter what.

Alternative energy sources have some great applications. But as a nation, we can gain only small amounts of electricity from burning wood chips and other material we can grow. And we can build only a few more hydroelectric dams in the West, because the best sites have already been used.

On the upside, thousands of wind turbines are springing up and researchers are working hard to improve solar-electric power possibilities.

But even when it's dark and calm outside—when solar and wind cannot help us—we want lights, coffee pots, and all the equipment in the ER to spring to life at the flick of a switch.

Those "must have" functions are part of what the folks who run the grid call baseload power, the electricity we always need. We need it on standard workdays, on major holidays, and on every day in between.

That brings us to the Big Three, the three viable sources of energy that are large enough to keep the power grid running across the nation no matter the weather. The Big Three are coal, natural gas, and nuclear power.

Coal is a favorite of geologists. The United States has a great deal of coal, and we could generate enormous amounts of electricity from it if we choose. We have more coal in the ground, by far, than the Saudis have petroleum remaining under their sands.

But coal is pretty dirty and dangerous stuff. Burning coal creates air pollution bad enough to contribute to the deaths of a couple thousand Americans each year (that's the government's estimate). And some of the pollution created by coal doesn't go into the air. It was ash left over from burning coal that created the sludge that inundated a Tennessee river not long ago.

And, as we all know, underground coalmining accidents grab headlines on occasion when miners are trapped or killed. On top of that, burning coal creates substantial carbon dioxide, a greenhouse gas.

The downsides of coal led us in recent years to build many natural gas power plants. But natural gas is an expensive fuel compared to King Coal. And natural gas, geologist T. Boone Pickens argues, should be saved for transportation, because it's a clean-burning fluid that can easily power cars and trucks. That's a good point, and we'll see if it gets traction in the American mind.

Our third option is nuclear power. Nuclear reactors make one fifth of the electricity we use each day. They've been helping to keep the grid up and running since the 1950s.

The accident at the Three Mile Island nuclear plant in the late 1970s clearly spooked us. The events resulted in no significant release of radiation, and no injuries or deaths at all, but fear of a catastrophe was intense at the time.

Still, when President Carter walked into the TMI reactor building, Americans realized the situation was under control. And from that day to this one, scores of nuclear reactors have been supporting the grid, every day and every night, no matter the weather.

If we look beyond our shores, of course, we can see other events that may make us decide we don't want to expand our nuclear power base. Most recently, the mega-quake and resulting tsunami in Japan have been a heartbreaking spectacle to behold—even across the broad Pacific Ocean.

We don't have ideal choices for energy. Our national baseload electrical options come down in practical terms to the Big Three. You can say we should not burn coal for electricity. You can argue we should not use a valuable fuel like natural gas to run power plants. And you can state emphatically your opposition to more nuclear power plants in the United States.

But remember, some of us desperately need our percolators to fire up— even on calm nights.

We cannot say no to all of the Big Three sources of power. It just ain't gonna happen.

Breaking Rock to Get at Oil and Gas

Where might you find the American who owns more petroleum and natural gas than any other citizen?

Alas, the answer isn't at my kitchen table. Instead, it's likely to be in the grassy fields of western North Dakota.

Harold Hamm, the son of Oklahoma sharecroppers, has risen in the world and become the savvy head of Continental Resources. That's a company you may not know but will hear more about in the years to come. It's pumping oil and natural gas in the northern Great Plains in rapidly increasing amounts, courtesy of a technology that hinges on breaking rocks that are thousands of feet under the surface of the Earth.

When water is under high pressure, it has the potential to crack rocks, and thereon hangs a tale of increasingly great importance to our national energy supply and also to certain environmental questions.

Deliberately cracking open rocks to help oil and natural gas move under our feet is called "fracking," a nickname for hydraulic fracturing. Companies like Hamm's drill down into the Earth and inject high-pressure water (with other ingredients) to create cracks in shale and then hold those cracks open. Other wells then pump out the fossil fuels that are liberated. Today about a third of the natural gas we produce in this country comes from fracking. The amount of crude oil brought to the surface with fracking is not terribly large, but it's growing very rapidly.

Fracking fluids may contain a variety of additives, and I do mean a wide variety. Ground-up walnut shells, of all things, salt, and citric acid (Vitamin C) may go down into the injection wells that create the cracks. More problematic can be benzene, which is a cancer-causing chemical, and a number of other trace compounds that may be added to the fracking system. The complex soup used in fracking helps to hold cracks open, to limit bacterial growth in the cracks, and to act a bit like soap—reducing surface tension—so that more fluids can flow into the extraction wells.

Environmentalists have gone to court to try to block some fracking operations. They fear the possibility of chemicals leaking into aquifers and from there into drinking water. Oil and natural gas companies counter that fracking is safe and that it gives at least some hope of increasing our energy independence for our transportation sector in the near term. And, indeed, from what I've read, American foreign oil imports are actually down as I write this due to the increasing supply of petroleum within our borders—largely associated with fracking.

Industry representatives also point out that there's a new commercial product that can be added to the water used in fracking, a product made only of chemicals used in the food industry. This, the argument goes, should lessen environmental concerns. But environmental groups appear to be highly suspicious of the product, perhaps in part simply because it's marketed by Halliburton.

There's no doubt that fracking has changed the domestic oil and natural gas industry, and in some respects this geologist doubts we'll ever really retreat from it. But it may evolve quite differently in separate parts of the country. In places like Pennsylvania and upstate New York, many people have deep concerns about fracking's potential for environmental impacts. In the rural and northern Great Plains, and in parts of the intermountain West, many more people may be concerned with the possibility of economic growth.

And that takes us back to North Dakota, where until fracking came along not a lot of the economic news was rosy. Now the oil workforce has grown more than three-fold in four years, and it may well head into the tens of thousands soon.

The jobs that fracking helps create are more varied than just drilling deep into the Earth. There isn't a big market for oil and natural gas in North Dakota, so a number of people are employed building the infrastructure to get the product of the Dakotas to where it's refined and used.

In some ways, the fracking debate is falling out along the traditional lines of environmental concerns versus economic growth. But I've got to hope that at least some cool heads will help us collectively consider what we want to do about a technology that could be of real use to us in some places and give us an increasing measure of energy independence.

Burning Our Own Fuels

British Petroleum's oil spill in the Gulf of Mexico has shown us just one of the downsides of petroleum production and use. Events like the spill make even a geologist like me turn to several questions about the future. Could Americans grow more of our own fuel—enough to run a number of our cars, trucks, and airplanes? And, quite importantly, could we do so without displacing food crops like corn?

Pretty much everybody from all sorts of political persuasions is interested in those issues. And the good news is that researchers—and farmers with a vision, too—are hard at work laboring on new uses of an ancient crop plant called camelina. Also known as "false flax," it's a wispy plant in the same group as mustard and canola.

There are two impressive things about camelina. Its seeds contain a lot of oil that's liberated by crushing—and the more oil the better from the fuel point of view. And beyond that, camelina can be grown on pretty lousy soil—either areas where no crops will grow well or at times that soil may otherwise be left fallow by farmers in dryland regions of our country.

Score two big ones for camelina.

Archeologists say that camelina has been grown by people for several thousand years. You can eat camelina oil, which has a lot of omega-3 fatty acids and antioxidants. Camelina was grown through the 1930s in eastern Europe where its oil was used in lamps and the meal left over after crushing the seeds was fed to livestock. In short, it has been a pretty durable friend of ours, as crops go—likely an original biofuel, you could say, and one that only recently fell fully out of use.

Today researchers are investigating bringing back camelina because it has some highly valuable properties for something much more modern than lamp oil. With some processing, camelina oil can be used in jet engines. It's a "drop in" fuel, meaning it functions just like traditional jet fuel. Blended with petroleum jet fuel, in fact, camelina has already been "test driven" in a number of jet planes all around the world.

Score a third big one for camelina.

"This ancient crop has a great deal of new potential," says Professor Ralph Cavalieri of Washington State University. "We think we are poised to make major contributions to biofuels with camelina."

But where can we grow this venerable crop in the United States that wouldn't displace food crops?

Here's an example. In eastern Montana and central Oregon and Washington, where the climate is semiarid, wheat is often grown on a rotating basis with fallow years. During the otherwise fallow times, camelina looks like a good crop. It can help hold soil in place compared to leaving it exposed to windstorms. And it could give farmers some income during years when they would have none otherwise.

It's true, of course, that when you burn camelina-based jet fuel—or corn-based ethanol or soybean-based biodiesel—you are liberating carbon dioxide, the well-known greenhouse gas. But that carbon is not ancient, like the carbon from coal or petroleum. It's part of the life cycle, if you will, of the recent history of the planet. The worry about global warming rests on our liberation of ancient carbon, the material that upsets the balance of what's in the atmosphere today.

There are many details yet to be worked out for camelina, and much work yet to do to see if it makes sense in economic terms on a large scale. But I'm hopeful oilseed crops will help us in several respects.

Still, it remains true that we cannot yet simply grow ourselves out of our dependencies on petroleum. By that I mean that the total biofuel power in the nation is a fraction of what we get from our main fluid fuels, petroleum and natural gas.

But every step we can take toward energy independence that doesn't limit our food production is clearly great news.

Hardy Bugs Are Eating Up the Mess

The BP oil disaster in the Gulf of Mexico has been plenty grim. I don't envy paymaster Kenneth Feinberg who has now taken over BP's $20 billion compensation fund. Feinberg is no stranger to trying to compensate those who have lost much, including the families affected by the events of September 11, 2001, and a somewhat lower-profile project to compensate victims of the shootings at Virginia Tech.

But the magnitude and complexity of the Gulf disaster is second to none. From what I can see, there's just no good way to distribute the money to those who have lost so much in different ways in various places.

Still, there's some good news. Since petroleum stopped flowing from the Deepwater Horizon site, oil at the surface has been harder and harder to find. And to some scientists, microorganisms look as if they may be poised to eat a lot of the underwater oil quite quickly.

According to ScienceNOW, the first peer-reviewed study to come out of the Gulf indicated the natural oil-loving "bugs" weren't doing as much as might be hoped. But the second one published looked at the microbes themselves and found much more reason to be optimistic.

Some bacteria in the seas have evolved over time to make a living on the seepage of oil that has always occurred naturally on the sea floor. The ScienceNOW website quotes NOAA marine ecologist Alan Mearns on the matter. "With all of the seepage, including the 40 to 50 million gallons a year that seep naturally into the Gulf, we'd have oceans covered with oil slicks if they weren't naturally degrading," Mearns said.

The bacteria in question consume hydrocarbons and oxygen, producing carbon dioxide as waste. The most recent study is based on samples of ocean water from a depth of more than 3,000 feet, both in an oil plume and outside it. The good news is that the study reports the concentration of the bugs inside the plume is twice as dense as in the water outside it. And the bugs in the plume have more of the genes needed to eat up the oil.

It's still an open question at exactly what rate the bugs are consuming the oil in the underwater plumes. Crude oil is a complex soup, made of many different types of molecules. The encouraging study concentrated on the alkanes, a type of molecule that's generally among the first to naturally degrade.

Another factor is that the petroleum mixes with water in complex ways. The term for the globs of oil in the water you've seen in the news is "emulsion." Emulsions are what results when two liquids that won't mix in chemical terms are physically blended together. You make an emulsion each time you shake up vinaigrette salad dressing, blending the oil and vinegar. Homogenized milk is a mix of fat and water so thoroughly emulsified that it's permanently mixed together, with no separation down the road, as in the vinaigrette case.

In the Gulf of Mexico, crude oil and seawater have been mixed together in a messy emulsion in the weeks following the disaster. What we don't know is how readily the emulsion globs will now break apart. We should keep our fingers crossed that they do so, because smaller and smaller bits of oil will give the bacteria a greater chance of being able to quickly eat up the crude.

One worry has been that, as the microbes in the Gulf go to work, they might use up the oxygen in the waters around them. That would create a "dead zone," like having too much fertilizer escape into waters with resultant algal blooms. Dead zones in the Gulf would be disastrous for fish that blundered into them. But the two scientific reports to date don't show that the oxygen has fallen to dangerous levels.

Considering the magnitude of the spill, the current scientific news from the Gulf is really quite good. Marine ecologist Mearns is hopeful that the oil plumes will be gone quickly.

"We're talking days to months," he said.

But I suspect Feinberg's headaches are only beginning.

Fuel Cells on the Horizon

What scientists call the "First Law of Thermodynamics" tells us that energy in our daily lives is neither created nor destroyed—only transformed. It's the chemical energy bound up in natural gas and oxygen that makes heat in your furnace, or electrical energy in my toaster that burns my bagels each morning.

But although energy is never destroyed, it certainly can be "wasted" or put to purposes we people just don't appreciate. My favorite example of that is an overheating pickup engine's radiator, boiling over vigorously on a mountain grade in the summer time.

My four cylinder "little rig," as we say in the rural West, gets about 28 mpg on the highway. Compared to what I get in my old pickup truck, I like them apples! But wouldn't it be grand if miles per gallon figures could be doubled? We would not be defeating the First Law, just using energy transformations more efficiently. And I'm glad to say I've seen a prototype of a device I think could do that—and economically, too. The little engine that could is called a fuel cell. There are several types of them; I'll just concentrate on two, first the one that's simpler but expensive, then the more complex but practical one that may revolutionize our world.

I've taught freshmen college students about the first one. It runs like a battery, you could say, but one that never needs charging because it uses fuel. Fuel cells have a couple clear advantages over batteries. Batteries are very heavy (come to my house and carry the boat battery inside in the fall or carry it back out in the spring if you doubt this statement). And batteries, for what they weigh, don't put out much oomph.

The simple kind of fuel cell draws in pure hydrogen gas to a negative plate, with oxygen fed at the same time along the positive plate of the device. There has to be a membrane and a chemical "soup" in the middle of the cell, but what matters for our purposes is that if you connect the negative to the positive parts you get an electrical current. You can run a light bulb, power a fan, or do any other form of work you care to with the resulting electricity. And the great thing about the fuel cell just described is that its only waste product is pure water.

But fuel cells like this are expensive. The catalyst inside them is made of platinum. And the hydrogen gas fuel is pretty scary stuff—hydrogen burns explosively with even the smallest spark. That's one reason why, although engineers have long been able to make desktop fuel cells for special applications, your car isn't running on one today.

But what if there were a second-generation fuel cell that ran not on hydrogen, but on mundane fuel like gasoline, diesel—or even coal or processed plant matter? All those materials could supply hydrogen to the fuel cell, even though not in pure form. And if they were consumed in the controlled manner fuel cells use to oxidize fuel, they would produce electricity, again like a battery.

At Washington State University Dr. Jeongmin Ahn and his students are developing these "solid oxide fuel cells." One of the great things about this next generation fuel cell is that is doesn't require platinum as a catalyst, using much cheaper nickel instead. Another feature of the fuel cells that knocks my socks off is that one model can run on waste heat. That's important, and here's why.

A standard car is about 25 percent efficient, meaning it uses 25 percent of the gas you put into it to get you from Town A to Town B—but it wastes 75 percent of the gasoline as heat! If a good measure of that waste can be fed into a fuel cell, it can help to power the car using an electrical motor—a new sort of hybrid vehicle.

"We can double the miles per gallon of cars using that technology," Ahn said to me. "Easily."

I surely like the sound of that.

Black Gold and Mileage

Crude oil is a mix of stinky chemicals. But smell as they may, the liquids in petroleum are vital to us.

An alert reader of these columns recently queried me about what we get out of a barrel of oil—and what those products mean in terms of miles actually driven on American highways.

Petroleum engineers "take apart" crude oil in refineries, separating it into different chemicals. Roughly speaking, a barrel of oil—which is a bit more than 40 gallons—gives us 20 gallons of gasoline, 10 gallons of diesel, and 5 gallons of jet fuel. (The stuff in jet fuel is also known to us old biddies as kerosene, but "jet fuel" certainly sounds more modern, doesn't it?) The remainder of the barrel is material like heavy oils and liquefied petroleum gases, the price of which you'll see in business reports as "LPG."

But what does a barrel of oil do on our roads? I'll answer that using the example of a recent model of the Jeep Grand Cherokee. I'm choosing that vehicle simply because it can be purchased with a gas engine, a diesel engine, or a "flex-fuel" engine that runs on either gasoline or 85 percent ethanol. I'll look only at city mileage—the lower end of the range you'd likely get with the vehicle. All my numbers are from a government website, based on the 4-wheel drive, 2008 model of the Grand Cherokee with an automatic transmission. I made certain choices about engine size in what follows. You can check variations on the figures or look up your own vehicle at www. fueleconomy.gov.

The Grand Cherokee gets 17 mpg in the city with a diesel engine and 13 mpg with a traditional gasoline engine. In other words, diesel gets you further down the road than gasoline does. For an individual person, that's all that may matter.

The downside of diesel includes engine "clatter" and challenging start-ups at bitter temperatures. The upside, beside better mileage, is considerably more basic power. That's important if you tow a big motorboat up and out of deep river canyons, as this Rock Doc has on summer evenings. (Can you say "boiling over"?)

If you want to think about what a single barrel of crude does on our city streets, the Grand Cherokee figures work out like this: A barrel of crude powers a diesel Grand Cherokee about 170 miles (because 17 mpg times 10 gallons of diesel is 170 miles) and that same barrel also powers a gasoline

Grand Cherokee about 260 miles (because 13 mpg times 20 gallons of gasoline is 260 miles). That's a total of 330 miles in the two vehicles from a single barrel of petroleum.

Moving heavy vehicles over city streets more than 300 miles is no mean feat, which is why we have a heavy national dependence on crude oil.

Perhaps you want to lessen our American addiction to foreign oil. If you buy a flex-fuel engine in your Grand Cherokee and run it on "E85," the fuel made of 85 percent ethanol and 15 percent gasoline, you get only 9 mpg in the city. That single digit mileage reflects the fact that ethanol doesn't contain nearly as much energy as gasoline or diesel.

Americans make ethanol from corn, and the process of growing, harvesting, and processing the corn uses a lot of energy from coal, from natural gas and, indeed, from petroleum. So ethanol is hardly free of fossil fuels, including foreign oil.

If you want to check out vehicles much more fuel-efficient than Grand Cherokees, the same basic link at www.fueleconomy.gov can take you to many choices. With fuel prices inching up, this geologist thinks that type of research will be increasingly wise.

The day is likely coming when many of us will commute to work in cars powered primarily by electrical energy, not liquid fuels. Plug-in hybrids of a couple different sorts will likely reach the average consumer in a year or two. Fully electric cars can be purchased now, but widespread use depends on next-generation batteries. Stay tuned for news.

Meanwhile, the stinky chemicals in a barrel of crude oil will remain central to our daily lives.

Harvesting Liquid Fuels from Trees

The goal of growing more fuel to meet our energy needs has strong appeal. Those of us who heat our houses with wood, of course, have been relying on trees for energy for a long time. But the more interesting task ahead may be to use trees for liquid fuels that can power vehicles.

There are two basic approaches that researchers are studying that could lead from trees to liquid fuels for transportation.

As high school biology teachers like to explain, the woody parts of plants are mainly made of two materials, cellulose and lignin. I think of cellulose as the tough stuff in plants, and because trees are large and strong, they each contain a substantial amount of cellulose.

For a simple rock-head like myself, cellulose is made of "hard nuggets" that are derived from sugar. One way to make liquid fuel out of cellulose is to use microbes that break down the "nuggets" to the sugars within them—material that then can be fermented into ethanol and burned as fuel. There are details yet to be worked out, but I think you'll soon see the day that genetically modified microbes will be used to convert cellulose to ethanol in a roughly economical way.

Lignin is the other material in trees. When you pick up a newspaper, you're holding a lot of lignin in your hands. Lignin turns yellow with time when it's exposed to air—a feature of old newspapers you've doubtless noticed.

Lignin gets in the way of processing cellulose for biofuel. Lignin and cellulose are both natural parts of the cells in trees, and we cannot simply eliminate lignin in a pine tree. But the day may come when we use biotechnology to manipulate the lignin-cellulose partnership to suit our purposes when the tree is harvested.

This is just one example of the fact that biotechnology—the art of manipulating the living world at the level of the gene and the cell—is the quiet revolution that's going on all around us. Biotechnology is not as obvious to consumers as the internet or cell phones, but biotech is reshaping the fundamentals of our world at a pace that will only increase.

Trees could also yield biofuels in quite a different way. Think for a moment of school lessons about rubber trees in the tropics. Many of us who live in northern climates vaguely remember being taught about rubber trees, but we don't know that other tropical trees produce sap that's similar to kerosene and diesel fuel.

That sounds odd, I know, but it's just a fact that some tropical trees naturally produce liquid hydrocarbon fuel. What's not to like about that?

The best "diesel tree" produces about 10 gallons per year of biodiesel fuel. And that's led some farmers in the tropical part of Australia to plant 20,000 of the diesel trees for later tapping.

Now, mind you, the quantities produced by diesel trees are small compared to what geologists rip out of the Earth in oil fields. Ten gallons per year is not exactly a flood of biblical proportions. Still, it's liquid fuel harvested directly from trees—quite a feat.

Here in the continental United States we don't have the climate for diesel trees. But some of our trees, like the Western red cedar, do have significant amounts of heartwood chemicals in them of interest for our energy concerns. Using biotechnology once again, we may be able to "improve" such trees from the point of view of biofuel production.

Professor Norman Lewis of Washington State University once gave me an overview of this whole realm of bioengineering. Everything from the lignin and cellulose divisions in plants to the possible hydrocarbon content of tree heartwoods is up for research these days.

"No matter how you look at our climate and energy situation," Lewis said, "there's a role for biofuels in some form. And I think in the next ten years you'll see some major advances."

If we can find ways to better utilize Earth's largest plants to produce more fuel, even us hard-core geologists will become tree huggers.

Dumping Gasoline, Going with Gas

B old visions have shaped American economic history at many different watersheds in our nation's history. T. Boone Pickens fits squarely in that tradition, applying the think-big approach to American energy needs. A geologist by training, Pickens has most recently been trading in oil and gas futures. He's also just published a book to tout his ideas about energy independence, and he's buying television ads to express his ideas.

Pickens' book, *The First Billion is the Hardest*, features information about the personal life of T. Boone Pickens. The last third of the volume, however, contains his message about a national energy plan.

Here's his sermon.

First, Pickens says we should use natural gas to power our cars and trucks. Second, we should use wind farms to generate the electricity we now make by burning natural gas in power plants. The second action will free up the fuel we'll need to run our cars and trucks. Problems solved!

Pickens is not a man troubled with details and parts of his vision are sketchy—as he freely admits. But he makes some good points.

It's not difficult to run cars and trucks on natural gas. Some American taxis run on natural gas, and most of the buses in Los Angeles are powered by it. If you've had a van ride at an airport with a company called SuperShuttle, you and your luggage have been ferried around by natural gas.

One version of the Honda Civic is manufactured to run on natural gas, and those vehicles are now being snapped up by economy-minded drivers in the United States. In some places in the country, natural gas costs only about a quarter of the price of gasoline, so there's no mystery about what motivates people to consider switching.

In Utah a number of people have started to get serious about running vehicles on natural gas. The governor has converted his SUV to run on it. (Cautionary note: Utah has unusually cheap natural gas. But no matter where you live, natural gas costs a good bit less than the equivalent amount of gasoline.)

By the way, burning natural gas is a heck of a lot cleaner than burning gasoline or diesel. Urban smog would be greatly reduced if the majority of Americans switched to cars powered by natural gas. What's not to like?

Pickens thinks we'll have a great deal more natural gas in the future as geologists learn to exploit a certain type of shale we've previously overlooked.

I'm less sure than Pickens that we are absolutely awash in natural gas, but Americans certainly have more of it than we do petroleum.

But before you put yourself on a waiting list with Honda or convert your present car to run on natural gas—a changeover that costs about $12,000—you need to think through some details.

Pickens is building natural gas refueling stations as rapidly as he can, but there are still none available throughout much of the nation.

You can, however, buy a small device that uses natural gas from a standard home line and compresses it to fill the tank of a vehicle. The refueling takes some time, but commuters could fill up as they sleep, with the clear reward of driving to work in the morning with vastly lower fuel costs.

An astute consumer may ask, what are the odds that natural gas prices will soon reach the price of gasoline?

Pickens writes, "Natural gas will always be cheaper than gasoline or diesel. I'll bet my ass on this one."

This geologist would hate to bet against Boone on anything relating to energy prices in the future.

The second part of the Pickens national plan is to develop thousands of wind farms to generate electricity—which will be needed if we move a lot of natural gas into transportation and thus come short on fueling the grid. Again, Pickens thinks big. He sees wind farms from Texas through the Great Plains to Canada. He's plowing hundreds of millions into wind-farm leases and wants to greatly expand the work. He demands that Congress take action now to make transmission lines from the Plains to our metropolitan areas.

Wind power is Pickens' biggest gamble ever, and he says it's his very best one.

Whether you agree with everything Pickens envisions or not, his current campaign is surely impressive work for an 80-year-old.

Geologists age well.

Thank goodness.

Gaining Energy from Splitting Water

On Saturdays this winter, when the snow hasn't been too deep in my backyard, I split a bit of wood for my woodstove. It's hard work, made doubly challenging by my arthritis. But as I address the woodpile, I like to dream of the day we may be able to set aside our axes and mauls and get energy from splitting water, rather than wood.

If I live to be an old lady, my dream may become concrete reality. It depends simply on some clever work by chemists and on the abundant energy we get each day from the sun.

The sun bathes the Earth with more power than we humans will ever need. The challenge about solar power isn't that we don't have enough, it's just how to fully use what we are given.

One route for throwing a harness over sunbeams is to learn to produce hydrogen fuel by "breaking apart" water. Splitting wood is hard work. Splitting water could be easy.

Water is made of hydrogen and oxygen. If we can split apart those two ingredients, we'd have a lot of hydrogen on our hands. Hydrogen gas burns well. We could use hydrogen as our basic fuel to accomplish everyday tasks ranging from heating the house to driving to work. And burning hydrogen rather than traditional fuels means we wouldn't be creating carbon dioxide—a real positive for climate concerns.

Scientific investigations into splitting water molecules and making hydrogen fuel are known under the phrase "artificial photosynthesis." That's not as formidable as it sounds. Indeed, it's mostly a matter of being as smart as vegetation.

When plants do their work of converting sunlight into chemical energy—the process of natural photosynthesis—part of what they accomplish involves taking water and breaking it apart. In plants, the hydrogen bits end up in sugars. Plants perform this trick via a complex organic molecule that contains manganese in its center.

It may be possible to create a man-made molecule that will do something similar. Research chemists like Jim Hurst at Washington State University are looking for ways to use sunlight to split water into hydrogen and oxygen in the presence of the special molecule.

Chemists have been able to create molecules that basically do what we're hoping for—but the molecules so far discovered in this arena are built around rare trace metals with names like ruthenium. The hope is to build molecules that can do all the work required and that will be created around common and cheap metals, like iron.

If chemists are successful, our world would change. We would have a source of hydrogen gas made directly by sunlight. We could use that hydrogen as a fuel, burning it in a flame or in a fuel cell (a device that creates electricity by combining hydrogen and oxygen gas in a controlled reaction).

Energy might become cheaper than dirt.

To be sure, there are some practical challenges, like storing hydrogen and shipping it. Because it's so flammable, we'd need a lot of safety systems around our hydrogen tanks and pipelines. But if we can address those concerns, and if Professor Hurst and his colleagues around the world are successful, we could run the machinery of our lives on hydrogen.

We could heat our homes, power our vehicles, and do everything else to which we've become accustomed, all with the energy we would harvest from hydrogen, created for us by sunlight. And because we'd be producing only pure water as the output from our tailpipes or chimneys, air quality would improve tremendously over what we've had while burning fossil fuels.

In short, there's a lot to like about the possibility of abundant and economical hydrogen.

Some cautious souls will say that the chance of harnessing sunbeams by splitting water is only that—a chance. That's true, but it's also true that chemists require only a tiny amount of funding to work in their labs, looking for possible molecules that could transform water and sunlight into hydrogen fuel.

And the payoff could transform our world.

Perhaps I'd miss heavy labor in the backyard with my ax and maul. But splitting water rather than wood looks better and better as I grow older.

Running Your Car on Air Alone

Kids delight in blowing up a balloon and letting it go. The air inside is under mild pressure, and when a youngster lets go of the neck of the balloon, air rushes outward. The escaping air propels the balloon forward like an erratic jet.

Remarkably enough, a car powered by the same energy source—compressed air—may be coming to a road near you. At least one innovative auto company is investing in a small "air car," as these vehicles are known. Air cars have some wonderful advantages compared to our traditional internal combustion engine—like the complete absence of air pollution coming from a tailpipe.

The idea of an air car is not so far-fetched as it may sound. Your commuter car, my 1987 pickup, and a farmer's diesel tractor actually all run on a broadly similar idea.

Work with me for a moment, and I'll explain.

The internal combustion engines common around us look like they are powered by heat from burning fuel. But all the heat actually does is to increase the pressure of gases in the engine's cylinder. It's the high pressure that pushes on the piston. Then the piston's motion powers the vehicle.

The heat isn't crucial; it's the pressure inside the cylinder that's the key.

Now imagine you could simply add highly compressed air into a car's cylinder to drive the piston. You wouldn't need heat, so there would be no need for gasoline or diesel fuel. And you could drive all day with no stinky fumes coming out your tailpipe.

For several decades, engineers have tinkered with the possibility of using air under high pressure to power the pistons of automobiles. The system can be made to work, especially if the air is under extreme pressures.

The pressures that are useful in a piston are generally much higher than those in a car tire. In scientific labs we often use very high pressure tanks, as do welders and others in particular industries. If you work near an enormous tank of this high-pressure variety, and it ruptures, your troubles are over. But I've never known that to happen.

If you've ever moved a high-pressure tank, you know they are heavy enough to give you a hernia. Indeed, the steel "fuel tank" of compressed air in old test vehicles was so heavy it created real trouble for the engineering goal of powering a car on air alone.

But much lighter-weight materials based on carbon fibers that can hold air at high pressure are now on the market. So visionaries are taking another look at the "air car," and carmakers overseas are exploring options of bringing such cars to market.

But, of course, there is the question of where the compressed air will come from. There's no such thing as a free thermodynamic lunch, and the cost of running the air car is partly the cost of energy to compress the air. As *Popular Mechanics* has pointed out, it's generally electrical energy that's used to compress air. So air vehicles are essentially electric cars using the compressed air as a way of storing energy.

On the positive side, pollution that's created generating the electricity used to compress air could be distant from our cities. That's a real plus. (Although we geologists are fond of the smell of spilled gasoline and the choking fumes of exhaust on a hot day, normal human beings prefer to avoid all that filth.)

A lot of innovation is on the table these days in the car world, with major manufacturers investigating better electric cars, hybrid vehicles, and natural gas vehicles.

These are tough economic times, but interesting, too, and some folks are going to take advantage of entirely new ways of doing things to help move us forward.

I'm for that.

SECTION 3
FOOD AND AGRICULTURE

The American farmer feeds not only U.S. citizens, but millions of other souls around the world. Our national success in agriculture has been directly shaped and aided in recent decades by research scientists. Some agricultural researchers breed better strains of crops, others work to find new ways to limit the damage done in fields by disease and insects. Very recently, agricultural engineers have been designing machines to harvest fruit in orchards, lessening the backbreaking and dangerous work currently done by workers on ladders.

Agriculture professionals provide us all with a lot of bang for the buck, doing everything from helping to stop potatoes from sprouting in our cupboards to breeding wheat that can cope with drought. All too often the public is blind to the important work done in agricultural research each year. Because I'm lucky enough to work with agricultural scientists each day, I'm in a position to give you a quick tour of a few highlights from their world.

Food and Agriculture

Killer Mushrooms on Your Plate

I t's a classic plot device of murder mysteries: an evil killer slips poisonous mushrooms into the frying pan of an unsuspecting victim who dies an agonizing death.

But in real life, poisonous fungi typically sicken and occasionally kill people for quite different reasons.

Recently I learned a lot about what can go wrong in the world of mushrooms from Dr. Denis Benjamin, a medical doctor who is also a fungi and poison expert.

Very young children (think toddlers) and dogs are two groups that manage to poison themselves during the warm portions of the year. What 3-year-olds and Fido have in common is that they are natural omnivores, moving around and putting most everything they find into their mouths. Often they have the sense to spit out odd-tasting objects with unfamiliar textures, but not always. Luckily, most mushrooms that grow in places like your backyard are not highly toxic, so a large majority of both toddlers and canines survive their experiments with fungi. But parents and dog owners sometimes get quite a scare when they see the objects of their love chewing blobs of fungal material.

Older kids can get into trouble because they dare one another to eat mushrooms they stumble across. Being brave in such games can lead to a stomachache or even serious medical problems.

Immigrants also run a real danger of eating the wrong mushrooms. While they may know safe mushrooms overseas, here in the United States some similar-looking fungi can be quite poisonous. A variation on this theme are mushroom pickers who hail from one part of the United States but use a mushroom field guide for another part of the country. That mistake is sometimes made even by experienced mushroom experts who fail to think through their methods.

In a related vein, it's worth emphasizing that matching a photo in a field guide or an internet source with what you pick isn't a good way of guarding your life. Many poisonous fungi are look-alikes for safe ones. Sometimes only microscopic differences separate the two—so don't go by photos as you decide what to eat for supper tonight.

Then there are the truly careless adults who end up each year in emergency rooms courtesy of mushrooms. It's no surprise that campers who are drinking heavily while spending time in the woods sometimes fry up what they pick among the trees. As the police blotter says about a variety of emergency situations, "alcohol was a factor."

Even sober, professional chefs make mistakes with mushrooms. The expensive Morel mushroom is a case in point. It must be cooked to decrease the toxin in its flesh. Unfortunately, from time to time, even professional chefs fail to remember this point, inadvertently poisoning their patrons with raw Morels in salads.

I once picked a whole hatful of what I hoped were Morels that had sprung up literally overnight next to the building where I worked. I'm no gourmet, so I knew that if what I had were really Morels, I wouldn't fully appreciate them. I therefore took them to a friend who really cares about food (and wine). He was delighted to get them, but I made it clear as I handed him the fungi that I took no moral responsibility for my gift. Still, overnight I had plenty of time to question my judgment in giving someone mushrooms I was in no position to truly identify.

My friend cooked and ate the mushrooms in the company of another gourmet the same day I picked them. The mushrooms were delicious, he told me the next morning, and I was relieved the meal had led to no ill effects.

That brings up an interesting question that Dr. Benjamin's presentation highlighted in my mind. Why do we always wonder, when we see a mushroom, if we (or our friends) could eat it and live to tell the tale? We don't feel that way about leaves of shrubbery, after all, but the question "Would BillyBob live if I fed him that?" seems to hover over all mushrooms we spy in the great outdoors.

Maybe we've all been reading too many murder mysteries.

Ready or Not

As events in Japan in March 2011 showed us, Big Ones really do happen. Richter 9 is about as large as they come, an event so enormous it takes away the breath of even a geologist like myself.

It's no comfort to think that quakes of that same general size are likely along the western boundaries of the Lower 48 and also in the region where Missouri, Illinois, Kentucky, and Tennessee come together. Even South Carolina is a hot spot of seismic risk. In short, major quakes all across the United States simply must be expected.

And there are other "big ones" too. As we see during the spring, flooding is most unfortunately a natural part of our world. Summer brings tornadoes, and the late summer is prime hurricane season. Beyond all that, electrical outages are disasters that sometimes shape the man-made landscape on which we depend each day.

Partly because I'm a geologist and know a bit about seismic Big Ones, I've done some work to get my household ready for emergencies. But recently I put myself to the test by comparing my paltry efforts with those of some of my Mormon neighbors. A local Latter-day Saints congregation was kind enough to educate me at an emergency preparedness fair they held for the public.

While I won't follow all the LDS advice—nor were the Mormons saying I should—I learned some pointers I will put into practice as time and my household budget permit.

There are many daunting questions to consider when it comes to emergency preparedness. To start simply, would you have water available if your tap water supply was interrupted? Just FYI, a gallon of water per person per day for two weeks is considered a standard by the American Red Cross. (My 5-gallon water jug in the basement surely doesn't look like much, does it?) My Mormon mentors also recommend having water filters or purification tablets on hand.

Could you heat your home next winter for a few days without electricity? Remember, a forced-air natural gas furnace won't run without electrical power. (On the question of emergency heat, I'm okay because I put a woodstove in the front room when I bought my house. Breathe a sigh of relief on that front.)

If the electricity fails, could you boil water to cook pasta or rice—and most importantly of all from my point of view, make coffee? (In the winter

I could use my wood-fueled heat-stove to boil water. In the summer I'd be coaxing water to a slow boil on my tiny camping stove. That would get old fast, for sure.)

If the grid failed, would you have any lights? (I've got two kerosene lamps, but they aren't bright. Somewhere I've got that LED light I used to wear strapped to my forehead for walking the dogs at night along the river. But where did that go?)

Here's another biggie: How long could you feed your household with supplies on hand? If the grid goes out, you can eat from your fridge the first day and from your freezer for perhaps two days thereafter. Then it's on to your shelf-stable supplies. But some of them, of course, will require cooking and hot water, which brings up earlier questions.

If you want to know the gold standard of preparedness when it comes to food, be advised the goal of many Mormon households is to have two years of staples in storage. But one reason for that high figure, I'm relieved to report, is that LDSers suspect they'd have to feed some of us heretics if a mega-disaster strikes.

So, if you do nothing else to get ready for an emergency, my advice is that you cultivate friendships with good Mormons.

Even if you never strive for the two-year standard of food stores, it can be useful to upgrade what you have on hand. A good pantry can be both economical and help buffer your household from a furlough at work or a cut in pay. And a well-stocked pantry really can buy you some peace of mind.

Having plentiful food stores on hand could also help make possible some extra donations to your local food bank when it's in special need. That, at the end of the day, might be the sweetest part of preparedness.

Ears of Corn

It's a world-famous grass, and crucial to our bellies. It's called *Zea mays* by botanists; rock-heads like me call it corn. Compared to many plants, it's excellent at tolerating drought and heat—almost in a class by itself in that regard.

Corn is obviously at the heart of corn muffins and tortillas, but you likely eat more corn in the form of corn-based sweetener in "junk food" and sweet drinks than directly as corn meal. Processed corn also gives us corn oil and corn coatings that are used in packaged foods.

But it might surprise you to learn that if you went to a standard supermarket, did your shopping, and then had eggs for breakfast, chicken salad for lunch, and beef for dinner, you would be, very likely, essentially eating corn at each and every meal. That's because many a chicken these days eats corn in its feed and some steers in feedlots gulp down corn on their journey to becoming bigger.

In short, and via many different paths, the 21st century American diet is built on corn.

You don't have to trace grain through the food supply to prove that statement. We can demonstrate it by taking a chemical analysis of our bodies.

Here's why the lab test works:

Zea mays is quite different from other grasses like wheat. It has a different way of capturing carbon dioxide from the air around it.

Now, as it happens, there are several kinds of carbon atoms in this world. They are all carbon—but they have slightly different weights (called isotopes in the trade). The carbon in corn has a different ratio of isotopes in it than the carbon in wheat because of the differences in the way the two plants "breathe in" carbon dioxide from the air.

So wheat, in this sense, really is different from corn, and a human body made of eating wheat is ever so slightly different from a human body made of eating corn. We literally are what we eat (which makes me a walking blob of peanut butter, but that's another story).

There was a day long ago when we grew a lot of wheat in the Midwest. But we switched to growing corn. The reason is that—using industrial farming methods—it happens that corn can be grown in the American Midwest in great abundance.

It's an accident, if you will, that *Zea mays* does so very, very well in our Midwest. But flourish it does, and with fertilizers derived largely from fossil fuels, we can grow prodigious amounts of corn.

Part of the complex story of King Corn is sketched in a book by Michael Pollan called *The Omnivore's Dilemma*. If you are looking for something to give you one person's view of modern agriculture, the book can be fun. I warn you, however, it's quite an opinionated piece. I'd say it's the kind of book that starts a discussion—not finishes it.

One point to always remember is that the American farmer has fed millions and millions of people both here and overseas. Part of the recipe for that accomplishment has been the success of *Zea mays* right here in the middle of our continent.

We've built a lot on an unusual grass—and been inventive in all that we do with its by-products. The question now is how much we like the current system, with its advantages and drawbacks, and whether we really want to make changes toward more diversification in our agricultural base. To do that, we all have to think through how we'd like to cook and eat, and what we want to pay for our meals.

The issues are so complex I'm certainly glad it's not up to geologists to decide them. But all of us, as citizens and consumers, determine such matters. So if you are looking for some post-dinner armchair travels through the food supply on long evenings, Pollan's book is one way to begin.

Eggs from Near and Far

To me, a good breakfast involves an egg or even two. That dose of protein, I think, helps me sustain useful work until noon or even beyond the lunch hour if need be.

Like me, you probably often have a dozen eggs on your grocery list. And when you wake up bleary-eyed on a Saturday morning, you face the choice of how you will buy those eggs.

In some parts of the country, there are three choices for procuring eggs. You can buy them at a supermarket, at a local farmers market, or directly from a local farm. If you want to support small farms—for any reason—then the second or third choice will be yours. But what if you care most intensely about what have been called "food miles" and how much energy is used bringing the food from the farm to your doorstep?

Food miles are the number of miles that food has traveled to reach you. It seems intuitively obvious that the lower the number of food miles, the less energy you are causing to be used for your groceries. It's better to buy food produced near you than food grown across the country, right?

Sadly, intuition does not always agree with reason and arithmetic.

Jude Capper of the Animal Sciences department at Washington State University recently took me through the example of buying eggs from the three sources mentioned above. The numbers that follow are just an example—your numbers would vary.

Let's say it's 1.5 miles from a house to the supermarket, 7 miles from that house to the farmers market, and 27 miles from that same house to a local poultry farm that will sell to the public. (Those numbers actually fit my situation pretty well, although they were chosen by Capper for another location.)

Now let's think of the food miles of the eggs themselves. In the case of the supermarket, Capper's example has them coming from 800 miles away in an 18-wheeler. Add the 1.5 miles for a person to get to the store and that's 801.5 miles of total driving around before the consumer first picks up the eggs.

"Obviously, on the first analysis, the food miles for the supermarket example are looking grim," says Capper with a laugh.

Even if the semitruck hauls other goods (like apples) back to where it came from, there's a lot of traveling involved to get eggs and produce to us.

The farmers market example and the local poultry farm case do involve less traveling for each egg. But there are other issues we want to consider since our real concern likely isn't actually food miles, but how much energy is consumed getting the eggs from the chickens to our frying pans.

Here are two important facts. Let's say the farmers market eggs get to their sales booth via a pickup truck, and I go back and forth to where I buy my eggs in a car.

I know it may not seem like it, but 18-wheelers are really quite fuel-efficient compared to pickups and cars when you consider all that they haul. Capper tells me they typically get about 5.4 miles on a gallon of diesel (plus, for a refrigerated truck capable of carrying eggs, they burn half a gallon of fuel per hour to keep everything cool). But the trucks move up to 23,400 dozen eggs!

Capper showed me the arithmetic that clearly shows the most energy efficient way for me to buy eggs for my household is to go to the supermarket, essentially relying on that highly efficient 18-wheeler rather than my less efficient car or the farmer's truck. And that's not even considering the notion that I'll likely go to the supermarket anyway, to buy laundry detergent, light bulbs, toothpaste, and bottles of eyedrops. (What can I say, I swim a lot.)

There are other reasons to buy locally produced eggs, Capper is quick to point out. You might want to support local agriculture, or you might prefer the taste of eggs from alternative systems. But if energy conservation is your primary concern in what groceries you buy, it pays to reason and go with the numbers rather than following your gut.

Giving You More Time
to Eat Your Potatoes

One of my mother's friends was raised decades ago on a few acres at the end of a gravel road in Idaho. As she puts it, her family's basic challenge was eating what the land produced before other critters did. In other words, it was useful to consume the eggs from the henhouse before something furry got to them.

Those images sometimes come to my mind when I look at a bag of potatoes in the grocery store. Potatoes are relatively cheap and nutritious food, a good source of the basic food energy that carbs give us, as well as a fine source of Vitamin C, potassium, and iron. They are also so easy to cook even I can manage the task.

But when I was young a lot of potatoes were lost to sprouting. As soon as a spud starts growing sprouts, its days are numbered. The sprouting hastens deterioration and the overall quality of the potato starts to drop off.

Farmers used to lose some of their crop to sprouting before they could sell their harvest, and consumers lost sacks of spuds stored in home cupboards to the same problem.

It wasn't that the critters of the woods invaded our kitchens to eat our potatoes. It was that the spuds themselves "ate" the starch stored in their pudgy brown bodies to fuel their growth.

In the 20th century farmers started to treat their potatoes with chemicals to delay the sprouting process. The chemicals helped, but they were artificial compounds, not natural to the food chain.

It's at this point that new research enters the picture. At agricultural universities across the nation, scientists are always at work to improve the lot of farmers and consumers alike. The goal is to get more food of higher quality out of the land, to do so as sustainably as possible, and to help consumers as we make a wide variety of choices about how we want to eat.

When Professor Richard Knowles of Washington State University saw that potatoes treated in an experiment with a naturally occurring compound didn't sprout—even when they were planted in soil—he knew he had discovered something akin to a magic wand for the sprouting problem. The compound in question is related to what wafts into the air when you mow your lawn. It gives you that "fresh cut grass" smell. In other words, the compound is fully natural, one that plants themselves produce.

The new class of potato-sprout inhibitors is being developed for commercial application by the American Vanguard Corporation as SmartBlock. It's because of advances like these that the big bag of spuds you carry home today from the grocery store will likely remain pretty cheap in the years to come. And as consumers we'll be able to directly see the good effects of Knowles' research, because potatoes will more likely make it from the store all the way to your plate rather than sprouting and growing furiously in the cupboard.

Most of the agricultural scientists I know are not the kind of people to loudly sing their own praises. And a lot of Americans today don't understand all the work that goes into growing high-quality food and getting it successfully to the grocery store. Those two factors combine to make too many of us unaware of all the good work done in agricultural research.

Lots of agricultural scientists are able to do their research work because of our tax dollars. At the federal level, agricultural research is supported by a part of the Department of Agriculture called the National Institute of Food and Agriculture (NIFA). The budget for NIFA was cut in a recent deal ironed out between Republicans and Democrats in D.C. Maybe that was reasonable given the economic situation, but because I'm lucky enough to see agricultural scientists at work around me in my job each day, I understand just how much bang we get for each buck put into their research.

If we choose to complain about the costs of agricultural research, let's at least not do it with our mouths full.

Is It Bread or Is It Cheese?

We Scandinavians have several strange customs, including our hallmark fish dish, which is cod that's been processed for days in caustic lye. This truly imaginative creation is known as "lutefisk," which means lye-fish.

Lutefisk is a jelly-like material with a pH so high it must be soaked in water for days before you stand a prayer of being able to choke it down. Even after the lutefisk has soaked a week, however, with daily changing of the water bath, never let silver touch the lutefisk, or the silver will be permanently ruined. And, as a final warning, if lutefisk remains on a pot or dish overnight, it's all but impossible to ever remove.

Would you like a second helping?

Although taste and smell are what we think about first when it comes to cooking, the texture of food is also a key factor in how we experience the process of eating. From that point of view, the interesting thing about lutefisk is the transformation of fish-texture to jelly-texture.

And that leads me to explain that Scandinavians have contributed another, quite different, food to the world, one that has some potential to be useful for certain people with special dietary needs.

Here's the story.

Some people become profoundly ill if they eat wheat. Folks with celiac disease have an autoimmune disorder of the small intestine sparked by gluten in grains. If you have the disease but avoid gluten, you're fine. If you eat gluten-bearing wheat, you become sick—sometimes very, very sick.

But if you're fond of wheat products, you know how discouraging it would be to give up the chewiness of pasta or the light texture of bread.

Enter a food science technician, one Michael Costello by name, at Washington State University. Costello has the bright idea to borrow a traditional Scandinavian food—a good one let me quickly add—and retool it to help replace pasta and bread with a gluten-free substitute.

Finns long ago created a cheese called juustoleipa, meaning bread-cheese. Some say the best bread-cheese is made from reindeer milk, but these days cow milk is the standard ingredient. Juustoleipa is a fresh cheese, meaning that it isn't aged.

Unlike most cheeses, the milk used for juustoleipa isn't fermented. The cheese-maker just curdles the milk and then bakes or grills it to give it a crispy crust. That unusual approach to cheese-making is possible because, unlike virtually all cheeses you have known, juustoleipa doesn't melt when heated.

Swedes like the cheese with coffee (true, we like everything with coffee, but this cheese in particular). The custom is to put a few pieces in the bottom of a cup, pour your coffee over them, and enjoy it as what we cleverly call "coffee cheese." The cubes stay firm and chewy and don't melt, hence the idea behind the custom.

Juustoleipa can become bread-like when it's prepared in an oven—air gets worked into the cheese as tiny bubbles as it bakes. So, depending on how the cheese is prepared, it can be chewy and pasta-like or chewy but more bread-like. Because it's really just a fresh cheese, it has a mild flavor that doesn't detract from any prepared dish. And, of course, cheese doesn't trigger gluten intolerance.

Costello teaches cheese-making techniques and has whipped up juustoleipa from time to time for years. But lately he had the idea of using it in ways to help those with celiac disease. He has made several dishes to test how the juustoleipa might substitute for wheat products in prepared meals. His most successful concoction is probably his wheat-free lasagna. The juustoleipa, sliced thin, takes the place of pasta—it's good, chewy, and the richness and oils of cheese is to be expected in lasagna. Actually, I like Costello's juustoleipa "bread," too. It's a bit like eating bread that has been spread with nicely browned butter.

Perhaps it's obvious, but we're not talking low-cal meals. A full slice of juustoleipa "bread" will stave off hunger for a good, long while.

But it's sure better than lutefisk.

A New Way of Making Supper

I take a personal interest in pickup trucks that can shut down half their cylinders to get better gas mileage when conditions permit. And I've studied the mechanics of hybrid cars that save braking energy to help power your vehicle a bit later in your journey.

Efficiency fascinates me.

But the efficiency of engines, as important as it is, pales in global significance to the basic efficiency of one piece of the living world.

Enormous business opportunities and satisfying meals both hinge on our ability to increase the efficiency of crops. And the good news is that researchers are working hard to greatly increase the efficiency of the most important staple crop the world has ever known.

Rice provides more than a fifth of all the calories humans consume. It is more important to stomachs worldwide than wheat, corn, rye, or even Big Macs. In short, more than 3 billion people depend on rice as their main source of food, including most of the world's poor.

But it happens that the basic way rice converts sunlight to carbohydrates is not efficient. In fact, rice is about as inefficient in the plant world as my 1987 pickup is compared to the new trucks that can shut down half their cylinders.

The rubber meets the road in the world of plants as carbon in the air is taken up and transformed into carbohydrates via photosynthesis. The source of carbon for plants is carbon dioxide gas. Yes, that's the famous greenhouse gas you've heard about, but it's also a natural component of every breeze you've ever felt, one that's crucial to all the plants around us. And from the point of view of plants, there isn't nearly enough carbon dioxide in the air.

As it happens, rice photosynthesis converts carbon dioxide into food in a pretty inefficient manner. In contrast, corn is a great example of a plant that makes much more efficient use of limited resources—including moisture in the soil and carbon dioxide in the air. Corn's internal processes are up to 50% more efficient than those of rice.

Plant scientists call corn a "C4" plant compared to the "C3" status of rice. Don't sweat the names—I mention them only because you may see them on the business pages of your newspaper. But focus on this important point: corn can actively accumulate relatively scant carbon dioxide in a way that supercharges its photosynthesis engine. So corn powers right by rice on the crucial racetrack of converting sunlight into food.

A number of research scientists now have projects underway to help create a new strain of rice that runs on corn's internal C4 biochemistry. It's like putting a modern and efficient truck engine into my 1987 pickup. This kind of highly promising work is part of the emerging realm of biotechnology.

It's not as easy to see the biotechnology in the world around us as it is to see the Internet and cell phones that clamor for our attention. But re-engineering the living world has potentially even more positive impact than improving electronics and engines.

Professor Gerry Edwards of Washington State University is one scientist dedicated to understanding the fundamentals of photosynthesis and the possibility of implanting the basic C4 engine into a C3 crop like rice. Under the leadership of the International Rice Research Institute in the Philippines, and through a grant funded by the Bill and Melinda Gates Foundation, Edwards is part of an international team at work on the project.

"C4 rice could yield more harvest on the same amount of land with increased efficiency in use of sunlight, fertilizer, and water. It's especially important considering global warming and water shortages," Edwards said to me recently. "C4 rice will require some years to achieve, but it's definitely worth the effort."

American researchers have always excelled at this kind of fundamental science, work that can make great leaps with practical benefits of consequence to billions of people. Research in science and engineering has been an economic powerhouse for us, and I'm sure it will contribute substantially to our recovery from this recession.

Pass the rice, please, and keep the good science coming.

Better than the Can

If you are ever cut off from supermarkets and electricity due to a natural disaster (or because like some of us idiots you choose to go camping), you will be especially interested in this news. And even if your only interest in daily life is eating well, read on for the glad tidings that are coming about how we will soon improve the way we process and store food in this country.

For about 200 years we've canned food in much the same way, putting it in cans (hence the name) and heating it under pressure for long periods. The heat both cooks the food and kills the bad little critters in the food. But heating the food for a long time is the reason why canned green beans are not like freshly cooked green beans and why canned salmon and chicken just are not like their freshly cooked equivalents. Canned food is better than nothing—because it keeps, because it's safe, and because some of us actually like mushy baked beans from cans. But compared to fresh food, most canned food runs a distant second best.

Many food companies, food scientists, and a variety of engineers have tried their hand in recent decades at finding a way to heat food and its container much more quickly and effectively. The goal has been to dramatically lower high temperature cooking times so that the taste, texture, and nutrition of the fresh food can more closely be preserved. Alas, most of the past efforts have failed.

But the good news is that, on the campus of Washington State University, Dr. Juming Tang is making great strides in bringing to your grocery store exactly what the doctor ordered. The food's safety has been demonstrated (even the worst "bugs" are killed). It gives us better taste, texture, and appearance (so people prefer the food). All this means it's what is known as a "shelf stable" meal that will taste and look much more like freshly cooked food, not like a traditional dollop of canned spinach oozing on your plate.

Tang and his team recently got Food and Drug Administration approval for one process, a crucial first step in cooperating with a consortium of companies and entities to bring the basic technology to market and then bring consumer products to a store shelf near you.

"This is the 21st century approach," Tang said in his lab. "In ten years, I believe that most companies that currently produce shelf-stable or refrigerated meals will use this technology."

The new processing technique depends on hot water and long-wave microwaves. (Dr. Tang's microwave isn't like the one you've got at home; his fills a very large room.) The microwaves focus energy onto the food. Careful and clever engineering, and the flowing hot water bath, keep the heat distributed, even at the corners, and fully up to temperature.

While a traditional canning plant has large boilers and is filled with humidity and wasted energy, this new approach can be much more efficient, cost effective, and environmentally friendly. It also provides a better work environment.

"The electricity for the microwave can come from wind or solar, and the heat for the hot water can come from the waste heat of the long-wave microwaves," Tang said.

The Army likes what it sees. Improving the shelf-stable foods that soldiers live on in the field is always one of its priorities because, as Napoleon once commented, armies march on their stomachs.

Major food companies are also paying a lot of attention to this project, and that's where you and I come in. If Tang and colleagues are successful— and I believe they will be—you can look for pouches or trays of food made from their process on the grocery shelves just a few years down the road.

Tang is the first to say he doesn't know how all the dollars and cents may work out. But the food companies that have seen his technology—and seen and tasted his products—are excited to explore this wholly new way of preserving food. They know which way the wind is blowing for cans.

Defeating Death

The gene for death has been isolated—and reversed—by scientists. Not a bad day's work, you might say, and a bright ray of light to cheer up us all. Sorry, it's not the death of human beings that's at issue. But it is a gene for death that's embedded in a plant on which we all directly depend each day. And that's good enough to encourage me about our collective prospects.

Wheat is the plant I have in mind, the king of cereal grains. It provides food for billions of us. Since the dawn of agriculture 10,000 years ago, farmers selected seeds for wheat plants that had bigger kernels that clung tightly together and could be easily harvested. For the past 100-plus years, scientific researchers have been adding to this 10,000-year history of selective breeding.

The sum of all that effort has allowed farmers to produce more and more wheat on the same amount of land, giving us more grain for bread, spaghetti, and my own personal favorite, huckleberry muffins.

But wheat dies in the field each July. So, to harvest wheat the next year, farmers must prepare the soil and replant. That's a lot of work, and it requires fuel and other inputs to accomplish.

If we could prevent wheat from dying each summer, it would grow indefinitely, like the grass in your backyard. Then we could harvest wheat without replanting. And this perennial wheat—with large and established root systems like grass—would be in our fields all year round, helping to hold our soils together through strong winds and hard rains. It's a clear "win-win," as the young people say.

The first step for creating perennial wheat is to change the gene that programs the plant to die each summer. Professors Steve Jones and Tim Murray of Washington State University are two scientists who have already accomplished that task. (Defeating death would make us geologists boast to the heavens, by the way, but both Murray and Jones are quiet and easygoing. Agricultural scientists are often like that.)

Death was defeated by crossbreeding wheat with one of its wild cousins.

"We're getting good at making wheat that doesn't die," said Jones. "Now we're struggling to make wheat smarter about how it continues to live."

In WSU test plots, the newly created perennial wheat faces several challenges. One is that after harvest in August, the scientists don't want the plant to perk up in fall rains and regrow another head of wheat.

"We need the plant to conserve its internal resources for the next spring," Jones said.

Ditto for getting the right genes into the wheat so that it doesn't perk up and grow like mad in a thaw at midwinter.

"There's a balance between turning off the genetic instructions for death and getting the right genes for living through the winter in ways that work well for us," Jones said.

Living ain't easy. Smart living is harder still. But you knew that.

There's another basic challenge for perennial wheat. It's more susceptible to viruses and certain other diseases. But the quiet agricultural scientists are confident about the work ahead.

"We'll get all of those issues addressed in time," Murray said.

Murray and Jones started work on perennial wheat about 10 years ago—a blink of an eye in the history of farming.

"But we're making fast progress," said Jones. "And there will be all sorts of advantages to having perennial wheat as a commercial crop where it makes sense."

It's not that all our wheat fields need be perennial. But there are places and conditions in which having perennial wheat as an option would be useful—even crucial for overall economics and productivity.

Science is a difficult discipline, and modern scientific research costs society a lot of money. But work like what Jones and Murray are doing has the potential to usher in a new day for farmers, conservationists, and the billions of us who eat wheat each day.

Defeating death is just frosting on the cake. A cake that's made, of course, from wheat.

A New Slice Out of the Apple

I often eat without thinking, either while listening to the news or writing. It's a poor habit for several reasons, one of which is my ever-growing waistline.

But the next time you bite into an apple, I implore you to take just a moment to really savor its taste, aroma, and texture. Those characteristics vary a lot between a Granny Smith and a Golden Delicious, or a McIntosh and a Braeburn. The variation is one reason apples are a delight.

Apples come from an ancestor tree that had small and acidic fruit. We don't know exactly how it was that people coaxed substantial improvement out of acid apples, but around 2000 B.C. people produced sweet apple stock. That was also when they began grafting branches from one tree onto another, a clever idea if ever there was one. Much later the Romans, who knew a good thing when they saw it, spread the sweet apple and grafting technology to many lands.

Now skip with me up to the present, because there's some big news about apples. In the parlance of biologists, the full genome of the apple has recently been described.

As it happens, scientists mapped the genome of the Golden Delicious apple. The Golden Delicious appeared on the scene as a "sport" in the soil of West Virginia, so it's as American as apple pie. It was propagated and sold starting in 1914. Other types of apples have come from it.

But what is a genome and why should you care about it? Some years back you likely heard that scientists described the human genome. But, as you've noticed, the people around you haven't improved one bit. So what's to be excited about with respect to our growing knowledge of genomes?

Let's focus on apples. The apple genome is the total of genetic information that governs how apples sprout, grow, and yield fruit. It's genetics that help determine firmness and taste of the flesh, as well as the aroma of the peel. Genetics also influence the resistance to disease an apple tree is likely to have.

If apples went through their whole life cycle every year like our grain crops, we could try crossing two related types and seeing what resulted about 12 months later. But apples are perennials, requiring years to take root, grow, and reproduce. On that ground alone, we could expect that manipulating apples to suit us by traditional methods would be slower than dealing with annual plants.

The day will soon come that knowing the apple genome will make us better at helping trees produce more apples. And apples that ripen at times and rates that are useful to us are another possibility, giving us the hope of more efficient harvests. Most importantly for us gluttons, apples could taste and smell like we want, giving us a way to directly blend a Granny Smith with a Gala.

As we zoom in and focus our knowledge of particular regions of the genome that do things like govern resistance to disease, we'll know more and more about what part of the genetic material of the apple we want to change. And we can then do exactly that in a lab, not in experimental orchards where we wait years to see the fruits of our labors.

An international team of researchers led by a group in Italy undertook the challenge of describing the apple genome to get a start on all this work. Professor Amit Dhingra of Washington State University was one person involved in the work, and he recently was kind enough to try to explain apple genetics to this rock-head.

Dhingra told me scientists estimate there are 900 genes related to disease in apples. As knowledge about those genes increases, scientists can help trigger useful mutations or sports, and then trees will have the ability—on their own—to stand up to natural diseases that sometimes plague them.

As a geologist, I'm not sure I followed the details of genomic research. But I do know that in a world of growing population, anything that can help increase, diversify, and secure the food supply is indeed good news.

SECTION 4
CLIMATE

Setting aside the politics embedded in many discussions of climate, some simple facts remain. First, climate is always in flux. We may yearn for what we remember of the weather in our childhood, but in fact what we grew up with was a simple snapshot of an ongoing and dynamic system that was changing. Second, there's quite a bit of evidence in nature about past climates. Geologists are one group of scientists who have studied the rich record of just how variable the weather has been. Here are a few thoughts about natural climate change I've had in recent years, as well as a plea that we put out unwanted coal fires around the world that produce meaningful amounts of carbon dioxide, thus adding to man-made climate impacts.

Climate

Playing with Jello and
Deducing Climate Change

I hope you played with your food when you were young. Perhaps you experimented at some point with pushing a drinking straw through Jello. If you twisted the straw as you removed it from your food, you could sometimes trap a column of gelatin in the straw. You then had the choice of either blowing the Jello at a sibling or, if your parents were at the table, gently squeezing the gelatin out of the straw onto your plate with your fingers.

Geologists take samples of ancient muck and mire in a way similar to kids playing with Jello. We bang pipes down into the soft Earth of lakebeds or peat bogs, pull them up, and push out narrow columns of mud inside. The muck is composed of many, many layers that go back in time. We geologists call this activity "coring," and although it's physically tough work, it's no more complex than jamming straws into Jello.

One reason geologists make cores of mud is that low spots on the Earth can record the climate of Earth's past. Evidence geologists get from coring lakebeds and peat bogs has taught us just how frequently both regional and global climate changes.

A Scandinavian geologist got the coring and climate story started. His name was Lennart von Post. He lived and worked around 1900, and he was the first geologist to carefully investigate what cores of muck could reveal about past climates.

Of all the places where geologists can core the Earth, our favorite spot is peat bogs. That's because peat is the first step in the long geologic process of producing coal, and geologists are inordinately fond of all fossil fuels. So it was quite natural that von Post started coring the ancient remains of plants and mud layers that make up the peat of southern Sweden.

Little fragments of twigs and leaves can be preserved in peat, and if you can identify the species of plant that produced such material you have your first clue about past climate in a region. Von Post went to work identifying such bits of old plants, but he also had the wit to look at the ancient mire through a microscope. What he discovered was that he could identify ancient pollen in the layers of peat he was cataloging.

Pollen is surprisingly sturdy stuff. It will remain intact for literally thousands of years, lying in a layer of muck, waiting for a geologist to come along, core it, and identify the plant that produced it.

If you have allergies, you know pollen is blown around on the slightest breeze. That's the basic fact that makes pollen much better than twigs or leaves for telling us past climate. Pollen reflects all the plants in a whole region.

If you know the identity of the whole range of plants in an area, you know pretty well what the climate must have been like, both in terms of temperature and precipitation. (Think of gardening "zones.") And once you've described the pollen from a core, you can make a carbon-14 date of a twig and assign a specific age to the climate you've been able to deduce.

Ancient pollen makes it crystal clear that climate varies again and again over whole regions on Earth. Just for example, in northern Europe where von Post first worked, there have been ten major climate intervals in the last 15,000 years. Each of these shifts was substantial.

The warmest era—when oak forests covered the lowland of Sweden—was what geologists call "the Optimum," the balmy times about 6,000 to 8,000 years ago. That era was much warmer than today.

Some of the great shifts in climate were global in scope, some were only regional. And just to give us all nightmares, some of the biggest shifts in temperature occurred in just 20 years or so—well within a single human lifetime.

Studying past climates demands strength in the field, patience in the lab, strong eyes for microscope work—and plenty of courage, too. The simple but brutal fact is that major and minor climate change is woven into the fabric of the Earth itself.

Reading the Record of the Tree Rings

Scientists have studied natural climate change for quite a while. Part of what we have learned about past climates comes from tree rings, and thereon hangs an interesting tale going back more than a century.

Flagstaff, Arizona, was a pretty small burg in the 1890s, without the street lamps of big cities Back East. It also has an elevation of 7,000 feet, making it well over a mile above sea level.

It was those two conditions that brought a young astronomer named A. E. Douglass to the area in 1894. He was commissioned to set up a new telescope by Percival Lowell.

Lowell was an amateur astronomer, fascinated by telescopic images of Mars that included long lines. The linear features seemed to run from the poles toward the middle of the Red Planet. Lowell speculated the features were Martian canals, dug by an advanced civilization to bring water from the poles to lands where it was becoming more scarce.

Unlike most of us, Lowell had the money to bankroll the investigation of his ideas. (In Lowell's family, when one generation died, it handed over real wealth to the next round or two of Lowells. When my grandmother died, I inherited a sweater. But I digress.)

Once Douglass got Lowell's telescope set up, he studied Mars alongside Lowell. But Douglass came to doubt Lowell's interpretation of what could be seen of Mars. In time, Lowell fired Douglass.

Douglass made a living in Flagstaff for a while as a justice of the peace and also by teaching. But he was a true scientist, and he kept his research interests alive as best he could. One of his interests was our Mr. Sun.

One thing astronomers had noted about the sun were dark patches, or sunspots, that sometimes could be seen on its face. The sunspots varied a lot in number over substantial periods of time, and they also appeared to go through a much faster 11-year cycle of smaller ups and downs.

Some people wondered if the sunspots created climate change on Earth. It seemed important to understand them better if they determined whether farmers would soon have good or bad yields. But studying the sunspot cycles was limited to the time people had been looking at the sun's face and keeping records about it.

Here's where some creative thinking comes into the story.

Douglass couldn't help but see the lumber industry working around Flagstaff, cutting down the big ponderosa pines of the area. He noticed that the width of the tree rings in the trees varied quite a bit. He wondered if the 11-year cycle of the sun had influenced climate in Arizona in a way that was recorded in the growth of the ponderosas.

Douglass went to work looking at the freshly cut trees. He measured the width of tree rings, the thick ones and the thin ones. He established clear patterns in the ponderosas of the area, with some distinctive "thick-thin" sequences in the rings. Of course, it was easy to count the rings back through time to learn in what specific year the trees had done well versus when they hadn't.

Then Douglas started to look at dead wood of the area. The outermost rings of some specimens of fallen trees sometimes matched up with the distinctive sequence of the living trees. When that happened he could count farther back in time with the older wood, and extend his thick-thin record keeping.

Next Douglass started to use old wood in the archeological sites of the Southwest, wood from Navajo hogans and even older structures. Eventually he and his colleagues, who had taken up his methods, had a good record of the thick-thin rings in the Southwest going back to the days of the most ancient ruins of the region.

What Douglass discovered in the tree rings were clear patterns showing how much climate could vary. There were years and decades of miserable tree growth, then long stretches of time in which the trees had flourished. Most interesting of all, it started to seem likely that climate change was a factor in what had brought early civilizations in the Southwest to their knees.

Tree rings are still being studied around the world. They give us one picture of how climate has varied—and it's not a comforting bedtime tale about a kindly Mother Nature.

When the Sahara Was Green

As schoolchildren know, dinosaurs lived in a much warmer world. Back in the era of T. Rex and Stegosaurus, dinos flourished all over the globe beside truly balmy seas.

Fantasize with me for a moment about going back in time to those days, when we could swim in the surf all year round and our December heating bills would simply evaporate in the warm breezes.

Earth's climate, clearly, has changed from what it was in the dinosaur era.

Recently a famous dino hunter came across evidence of much more modern climate change. Paul Sereno of the University of Chicago is a geologist who is expert at finding and unearthing dinosaur bones all around the world. (If you know an 8-year-old, ask about Sereno and you'll learn he's the cream of the crop in the realm of dinosaur discovery.)

In recent years Sereno has made a specialty of hunting dino fossils in the Sahara. And by chance, his crew stumbled across dozens and dozens of human skeletons in loose sands, lying above the dinosaur-rich solid rock below the dunes.

The serendipitous find led Sereno to unearth the bones of people who once lived in what's now the barren desert of Africa. The bones tell a story of early Stone Age humans from cultures of different times. They reflect two distinct burial styles and are associated with pots having quite different patterns. In other words, there are twin pieces of human history wrapped up in this one special spot in what's now an empty desert.

Here's the climate connection: the human bones are accompanied by bones of animals, including crocodiles, hippos, turtles, clams, and various fish. The great supply of water made evident by these bones has led to the nickname "Green Sahara" for the time of the Stone Age peoples.

Both archeological evidence and radiocarbon dating show the age of the younger culture as 6,000 years ago and the older one as 9,000 years ago. Just before that earlier time, Earth's climate had warmed up from the bitter cold of the Ice Age—the frigid period in which Canada, the upper Great Plains, the Midwest, and New England were all under thick glacial ice.

In Africa temperatures didn't climb much when the Ice Age ended, but weather patterns shifted. The Sahara suddenly got rain each year—lots and lots of rain. The humans Sereno found were buried in what was then a small peninsula, jutting into a local lake in which the crocs, clams, and fish lived.

It was, clearly, a good spot, one used by people for literally thousands of years during that wet period.

As a geologist I'm happy to report the people of the Green Sahara knew the value of rocks. They made stone tools out of an unusual volcanic rock found about 100 miles away, either traveling there to bring it back or trading with other people for the valuable material. We know they made stone tools from that rock beside the lake because of abundant rock flakes in the loose sands.

Sereno's work shows the Stone Age dead were buried with care. A woman reaches out her arms to two small children in one gravesite. Other individuals were buried with pots and even flowers.

But then the whole scene changed. The wet winds shifted, and by 3,500 years ago, the Green Sahara literally dried up. It was a climate upheaval more massive than anyone alive at the time could have understood. In short, the Sahara lurched back to its previous status as a desert. The water-loving creatures and the people of the area were soon replaced by arid sands.

In each chapter of Earth history, Mother Nature makes it obvious she likes a climate story with a truly vigorous plot. That's not to our liking, but such is the clear record geologists and archeologists have learned to read from all over the world.

Sea Changes Bigger than Tsunamis

In the mid-1800s there was a smart geologist with a face as sharp as flint who worked in the American Midwest. In those days, the "Midwest" was quite close to the frontier of the country. It took some guts and imagination to live out there, and I suspect Charles Whittlesey had both in abundance—for he clearly saw evidence of dramatic climate change in riverbanks and hillsides around him. For Whittlesey, the Ice Age was evident in almost every field and ridge.

Many geologists of the time were still skeptical of the new theory that Earth's climate could change at all. It wasn't easy to think that Mother Nature had once put the whole globe into a deep freeze, but my hero got onboard with the program early. He argued (correctly) that much of the upper part of our country had once been buried under thick, glacial ice, and he did so by pointing to specific pieces of evidence he could describe and draw.

That alone would make Whittlesey commendable in my book. He looked at good evidence, published as widely as he could at the time, and argued for his views.

But this is what really impresses me. Despite the fact Whittlesey was living and working in the Midwest—pretty far from the ocean—he had the insight to see that the massive glaciers of the past must have changed global sea level drastically.

Here's the picture:

During times of bitter cold in the past two million years, extensive ice sheets and major glaciers have formed in North America and Scandinavia. While those glaciers were draped on the land, they "locked up" a great deal of Earth's waters.

More and bigger glaciers meant lower and lower sea levels in the Ice Age, a point Whittlesey deduced early. One of his estimates put sea level of the Ice Age at around 300 feet lower than today, a value that stands up well to current scientific data.

With sea level hundreds of feet lower than it is now, brown bears (and people) could walk from Siberia to Alaska—and they evidently did so, spreading down into North America.

But climate naturally evolves on Earth, and in due time the Ice Age came to its end. When global temperatures shot upward into the warmth we enjoy in this epoch, the massive glaciers and ice sheet started to melt. Water that had been on land (as ice) flowed down into the seas. Ocean levels rose, and rose, and rose some more, an increase totaling hundreds of feet.

But oddly enough, that's not the end of the story. In some places—like Sweden and Hudson Bay—the land is rising out of the sea. In other words, in those places, *local* sea level has been falling even while global sea level has been rising.

In Scandinavia and Hudson Bay, the evidence the sea is falling compared to the land is the many old beaches that are high and dry on the land, well above current sea level. These "raised beaches" show us the land is moving upward even faster than global sea level has been rising. But there are not raised beaches like these everywhere on Earth, only in places where (interestingly enough) major glaciers used to lie.

Our hero had some insight on this issue, too. The ice sheets of the Ice Age were literally a couple miles thick and covered whole regions. When the Ice Age glaciers melted, their staggering weight was removed. Gradually, the land under the ice has moved upward—and it's still doing so.

The land in Hudson Bay and Scandinavia is headed higher and higher at a faster rate than global sea level. So local sea level where my herring-eating ancestors live in southern Sweden is dropping relative to land.

Climate change on Earth guarantees that sea level can go up a whole lot in some places and, at least relatively speaking, down in others. Those are the realities to which people have adapted for a long time and will doubtless have to do so in the future.

Like climate itself, the only constant we geologists can see when it comes to sea level is ceaseless change.

Ancient Poop Yields Clues

My favorite epoch in Earth history is the Ice Age, the time in which saber tooth tigers and giant mastodons roamed the world. The Ice Age ended 10,000 years ago when—quite abruptly—the bitter temperatures of the time gave way to our present, balmy epoch.

Natural history museums often have the skeletal remains of Ice Age mammals. They are enough to inspire awe in part because many of the species alive during that time were much bigger than modern animals. The Ice Age was a time of giant deer and moose, with a species of beaver as large as a modern black bear.

In short, it was a time not long ago with a climate and ecology so different it intrigues both geologists and school children.

Imagine my pleasure, then, when I recently got to hold a sample of 16,000-year-old woolly mammoth poop from the Ice Age. Talk about an intimate connection with the past!

Ancient poop is known to geologists as coprolite material. It can be truly fossilized as solid rock, or just preserved in glacial ice, permafrost, or dry caves. One geology department softball team I knew proudly named itself the "Coprolites." Such is geologic humor.

Ancient mammoths and other prehistoric animals were well defined by the phrase, "you are what you eat." And by studying what ancient animals ate—through their intestinal remains or their poop—we can learn about their diets and nutrition.

Bruce Davitt of the Department of Natural Resource Sciences at Washington State University has the habit of studying ancient and modern poop—and deducing from it important clues to animal habits and environments. In one case he looked at caribou dung found in a permanent snow patch in southern Yukon, outside of Whitehorse. Caribou haven't been present in the area since the 1800s, so obviously the dung was at least that old. But using radioactive-dating techniques, researchers found the dung was about 2,400 years old.

Enter Davitt and his special talents at examining poop.

"We looked at the material under the microscope and compared plant fragments in the fecal pellets to the plants of that environment," he said. "We determined the dung reflected caribou's spring or summer diet."

It's possible the caribou sought out the cool temperatures of the permanent ice patch in the warmth of the Arctic summer, either for the temperature change alone or to get away from the summertime bugs that are thicker than pea soup in that part of the world.

The permanent snowfield also yielded part of an arrow shaft which was dated at about 4,400 years old, making it a rare example of hunting technology from that era found in Canada.

Davitt has also worked on much more recent fecal material and deduced something crucial to the health of some langur monkeys in the Bronx Zoo.

As Davitt explains it, several monkeys in the Bronx had mysteriously developed cases of peritonitis. The zoo sent Davitt samples of the animals' intestinal products to check for what could be contributing to their life-threatening illness.

"The vet there at first thought it might be that the monkeys were eating rope that was in the exhibit," Davitt said. "But monkeys have a strong sense of social order. The higher-ups on the scene got to eat the monkey food the zoo provided while lower-downs waited nervously on the side."

That led the less fortunate monkeys to eat fronds of a plant called *Pandanus utilis* that was in the exhibit.

"We found the plant in the fecal material. There were microscopic crystals on the plant fibers, and those crystals were rubbing on the monkey's intestines," Davitt said.

That insight helped save the lives of the monkeys after the plant was removed from their environment.

"We're just a small part of much bigger teams, helping researchers elsewhere engaged in their studies of wildlife," Davitt said.

One simple piece of equipment in Davitt's lab is a kitchen blender. It's used to pulverize samples into a thin soup from which slides can be made for examination under the microscope.

"We call the product in the blender a crappé frappé," Davitt quipped.

Break the Glass, Douse the Flames

As everyone who watches the evening news knows, in the western United States wildfires and forest fires are common enough in the late summer. Young people work diligently on fire crews in the West, fighting one of nature's great forces. Out-of-control blazes in our National Forests are all but an annual event, with only the number and intensity of the fires varying from year to year.

Wildfires in grasslands often last perhaps a few days. Forest fires are generally measured in days to weeks. In other words, all such fires are relatively short-lived. And that, as my young friends would say, is a "clear positive."

A clear negative must be acknowledged for another type of uncontrolled fire, a kind that burns year round, not just for a season but, in fact, generally blazes for decades. These are the fires geologists know best, and it's time that others learned about them.

The fires I have in mind are made of burning coal. Such unwanted coal fires rage or smolder in the United States, South Africa, Australia, China, India, and beyond. They are burning in huge volumes in rural China and blazing in a district of India to such a great extent the flames from some surface coal fires are over 20 feet high. Here in the United States, they are burning in Pennsylvania, Kentucky, Colorado, and Wyoming as you read these words.

People with little experience of coal may think it should be a simple matter to put out a coal blaze. Buckets of water, some would guess, should quench the flames, like water on a charcoal grill. But, in fact, coal can burn very hot (when oxygen is available at the surface) or smolder slowly (when little air is around under the surface). And once a major coal seam is ignited underground or near the surface of the Earth, it's quite difficult to control. Putting out a significant coal blaze by hand is almost impossible.

The public might also be surprised to learn the total effect of these unwanted coal fires. Carbon dioxide production from unwanted coal fires around the world is absolutely enormous. And because coal underground burns incompletely to a variety of gases, there are also other greenhouse gases beyond carbon dioxide that are at issue, such as methane and carbon monoxide.

According to one technical paper from the Department of Energy I have studied, about 2 percent of all annual industrial global emissions of carbon dioxide come as a byproduct of unwanted coal fires in China alone. In other words, coal fires in China—which are legion because their mines are often hand-dug into near-surface coal seams, a practice going back to antiquity—are adding a couple of percent to our total global production of carbon dioxide from industrial sources like power plants and auto engines.

But I'm not picking on China. We've got coal fires burning here in America, including one near Laurel Run, Pennsylvania, that's burned since 1915. And perhaps you've heard the story of a hamlet called Centralia that's also located in coal country in Pennsylvania?

Centralia is basically unlivable today because of a coal blaze. The air is tainted with smoke and toxic gasses, and the ground itself collapses from time to time. It has these characteristics because, when the town's landfill trash was burned off in 1962, it lit a coal seam just under the trash pit. From that day to this one, the fire has been burning.

You may remember that at the end of the first Gulf War, Saddam Hussein's forces left Kuwait's oilfields ablaze when they retreated back to Iraq. It was said at the time that putting out the oilfield fires would take years. But in fact, due to the work of an American company and the best technology of the day, most of the fires were out within months.

Coal fires pose some different problems than oil fires, but it's time we turned up our sleeves to address the blazes that we most likely can douse. We would be helping the folks living near the fires today, as well as our posterity tomorrow.

Two Ways to Think Cool

Dogs pant with their tongues hanging out, young men sweat by the bucket, and aging geologists just fall over on our faces in the shade on a hot summer's day. But is there a way we could choose to cool the whole planet in a few decades if we really need to?

Global warming may or may not be our greatest problem in the 21st century. But if it is, there are two ways we could potentially lessen warming. One is exotic, but would address the worldwide problem of warmth. The other is more closely attuned to practical matters and common sense, and would help us limit the heat increase that's so prevalent in and around cities—where most Americans live.

Here's the global picture:

Astute readers of the news media know that each time there is a major volcanic eruption—like Mount Pinatubo in the Philippines in 1991 or Mount St. Helens in my own fair State of Washington in 1980—the Earth cools. Pinatubo, for example, cooled the whole world by about 1 degree Fahrenheit. The effect lasted about a year. In strong cases, the volcanic cooling impacts agriculture and generally decreases crop yields—which is why it shows up in the news and normal people, not just us rock-heads, hear and care about it.

Volcanoes act as year-long cooling agents primarily because of the sulfate in their eruptions, tiny bits of which make it up to Earth's stratosphere with the force of the blast. Humans also create sulfate particles in the air when we burn coal. That's because coal, especially the nice, cheap coal geologists love, has quite a high level of sulfur impurities in it. Our coal-based sulfate in the air sometimes stays much closer to the ground and contributes to acid rain just downwind of smokestacks, but some of it gets carried upward, too. In either case, it acts as a bit of sunscreen that lowers temperatures.

There are now a few serious proposals about doing more, not less, sulfate "pollution" of the upper atmosphere. If global warming becomes severe, the argument goes, we could launch sulfate into the stratosphere to mimic the natural volcanic effect and lower global temperatures. The work would have to be done each year, again and again, but all sorts of launching devices could be employed, including relatively cheap devices like airplanes, balloons, and perhaps even battleship guns.

I doubt it will ever come to that, partly because we could expect crop yields to drop. But I do think we could become smarter about a simple matter that would make our cities and suburbs cooler—and actually cool the planet just a bit as well.

Here's a simple fact: dark colored objects warm up a lot more in sunlight than light colored ones. And, as you know, many roofs are dark: gray-black or dark brown asphalt shingles are popular on houses, black tar on flat roofs, and my own personal favorite, dark slate on a few roofs of historic stonewall houses back East.

What all these roofs have in common is that they warm up greatly when they are bathed in sunlight. That makes for hot roofs, but also for hot air all around the roof during the summers—thus contributing to hot cities and suburbs from Memorial Day to Labor Day.

You can actually see the effect in weather reports during the summer. Temperatures in the country on scorching days in July are routinely lower than in the suburbs and cities. That occurs for several reasons, including that the country is full of plants that are pumping water up into their leaves where it evaporates (and us country bumpkins stay cool because of it). Meanwhile, cities and suburbs show mostly roofs, parking lots, and streets to the sun—and bake all afternoon long because of it, achieving higher and higher temperatures.

If summer heat turns out to be our greatest trouble, we might be well advised to think about light-colored roofing. You could say that each white roof is like one small bit of snow lingering on the Earth's surface all summer long, reflecting a great deal of light and energy—so, one roof at a time, we could actually help the whole Earth.

Three Bears Dance with Climate

An unusually large grizzly sat on top of a deer carcass outside of Cooke City, Montana, one fall not long ago. Locals got pictures of the great grizzly, estimated at 900 pounds, which evidently was guarding the deer by sitting on it.

"It's a bear in good fall condition, not an old behemoth," said Kevin Frey based on the photos. Frey is a bear management specialist for Montana's Fish, Wildlife, and Parks Department, and his opinion was reported by the *Billings Gazette*.

People around Cooke City had other grizzly encounters that same summer. A bear bit a camper inside a tent in July. The next month, another bear actually entered a home.

Geologists doing fieldwork encounter grizzlies from time to time. One colleague of mine took a .44 caliber revolver with him as "bear repellant" when he headed to fieldwork in Alaska. Alas, he put the weapon in his carry-on luggage and took it to the airport, managing to get himself arrested. (Geologists are not necessarily the sharpest tacks in the box, a fact I illustrate when I cannot find my bifocals even though they're on top my head exactly where I left them.)

The grizzly activity in Montana is an indication that the great bears have staged a partial comeback since their low point in the early 1970s.

Bear researcher Charles Robbins of Washington State University is one scientist who has watched that comeback closely. Robbins and his graduate students study grizzlies in the wild using GPS collars. The collars record where the bears have been over time. Researchers on the ground then backtrack the bears, looking for information about their behavior and feeding habits.

"Their scat tells us what the bears where eating," Robbins said.

Robbins also studies grizzlies year-round at WSU's bear research center, where captive bears live and reproduce. Interacting with hibernating grizzlies in the winter at WSU is quite different from what the good folks in Montana experience some summers.

"The bear's heart rate drops down to eight or nine beats per minute and two of the four chambers in the heart stop beating," Robbins said. "Even so, the bears can stand and walk. All of that could help us develop medicines for people with high blood pressure and congestive heart failure. And despite lying around for months at a time, bears don't lose muscle mass or bone density. That's quite a trick we could learn from."

Grizzlies face a number of challenges in this changing and uncertain world.

"Many grizzlies depend on salmon as a food source," Robbins said. "As ocean conditions change with global climate, there may be a real impact on bears."

Grizzlies are already veterans of climate change. They are an offshoot of their close cousins the coastal brown bear that migrated across the Bering Strait during the Ice Age, when the global sea level was lower because so much water was frozen into glacial ice.

Modern brown bears are similar to the ancient, immigrant bears, while grizzlies are their inland cousins. Polar bears are an offshoot of both the grizzlies and browns.

"Polar bears are extremely well adapted to life on the ice. Being white is just one small part of it," Robbins said.

A polar bear once pursued a geologist I heard about in Greenland. The fellow had no gun, but he quickly took off his coat and dropped it as he ran along shore. As the animal paused to sniff and maul the coat, the geologist was able to reach his boat, and the bear didn't pursue him into the water. (Occasionally even we geologists can think clearly.)

Robbins tells me that polar bears changed quite quickly during the Ice Age as they evolved from brown bears. And since coming onto the scene, they've proven themselves significantly adaptable to some climate change.

About 10,000 years ago, the bitter cold of the Ice Age gave way to warmth like the present. Polar bears survived, although their habitat must have decreased greatly. They also lived through the warmest part of the current epoch, a point called the "Optimum" that occurred several thousand years ago.

But it's unclear how polar bears may fare if climate warms above both recent temperatures and then potentially above even the Optimum. That distinct possibility has a number of wildlife biologists highly concerned.

The tale of the three bears is still being written around us.

SECTION 5
HUMAN HEALTH

O ne of the astonishing success stories of modern science has been everything related to the medical realm. Life expectancy has increased through the last century and into this one. Childbirth is no longer viewed as a dangerous business and, at least for most of us, if we exercise and eat right we are likely to live longer—and live better—than our grandparents.

Our cultural memory is flexible, and we easily adapt to good news. Thus we take for granted major interventions in our lives ranging from treatments with antibiotics to having our knees entirely replaced. Medical research can be highly applied, but some of the rapid progress we've made depends on more fundamental research undertaken by a wide range of scientists. Here are just a few pieces that show some dangers that remain in our environment and how scientists are researching a wide variety of topics related to our health.

Human Health

From Old Moms to
Plastic Water Bottles

Science is full of unexpected discoveries that crop up in ways and at times no one plans. Just as there can be a breakthrough in negotiating the end to a war that comes to everyone's surprise on a major holiday, or intense spurts in which a writer completes a book in little more than a week, there are times a scientist may feel awash in troubles but then see the world afresh as the facts fall into place in a new way.

An element of chance enters science, just as it does everyday life. But, as Louis Pasteur noted, chance favors the mind that's prepared to understand the world from a new angle.

Here's the story of a prepared mind and the discovery that had major implications for our plastic water bottles and other food packaging.

Professor Patricia Hunt is a genetics expert who investigates how and why older moms are more likely to give birth to babies with a number of birth defects. Many of us women don't know it, but the tiny eggs in our ovaries are actually as old as we are—we don't make them afresh as life goes along. And the older those egg cells are, the more likely they are to have certain problems. To be blunt, older moms run a greater risk of having babies with certain birth defects.

Scientists do study human moms, but we can also learn a good bit from seeing how and why the eggs inside a "mouse mom" change over time. When scientists study animals, we generally have one set of critters with problems and another group called the "control." The idea is to treat both groups in exactly the same way (same food, same bedding material, and for mice the same little exercise wheels and tiny televisions). Any differences between the groups should come about due to the one crucial difference that's being studied, say the presence of a particular gene or the age of the egg cell.

Normally, the control group shows boring results. It's the regular stuff of mouse life. But some years ago, Hunt found she had a puzzle on her hands. Quite suddenly, even the mice in the control group in her lab were having a lot of miscarriages. Hunt and her team had to retrace their steps and look for anything that would explain the change.

In the end, Hunt discovered that the mice's plastic water bottles had been cleaned at one point with a harsh detergent. That detergent apparently meant that some chemicals in the plastic—chemicals that are like the hormone estrogen—were moving from the bottles into the water. That small change created the big differences in the mouse mom's reproductive life.

The possible implications of Hunt's discovery for people are substantial. We use hard plastics every day. They come in different kinds, but some of them can add the estrogen-like chemical bisphenol-A into the liquid or food they contain.

In response to the discovery of the problems the mice had, millions of Americans have decided they want to limit their exposure to the chemicals that come out of hard plastics. Plastics advertised as "free of bisphenol-A" are one option, and you can see lots of them at the store. (My 40-year-old glass thermos purchased for 50 cents at the next-door-neighbor's yard sale works nicely for me, but that solution will be over the day I drop the Thermos.)

Thanks to Professor Hunt's work and her prepared mind, we've learned a lot about plastics in recent years, including some surprising discoveries that came along in unexpected ways. The fruits of science are not always what we can predict, but they make a valuable harvest—and they can change even basics in our kitchens and lunch boxes.

Pasteur was right in emphasizing we must prepare our minds—and I would add those of the next generation—so that we all can continue to make the discoveries that will propel our progress in the 21st century.

New Twist on Natural MRSA

Twice a week I slog through some reps at the gym to keep my core muscles and arms strong enough so I can do basic outdoor tasks. I assure you, my workout regime is pretty modest, based on small "ladies" weights—some of them even colored pink.

But whether you use pink weights or the contortion machines that defy description, gyms ask you to wipe down your equipment with alcohol when you're done. The main reason for that rule is a type of bacteria called staph, and a strain of staph called MRSA (that's pronounced 'mersa' by the medicos).

Here's the basic picture about staph.

Whether we work at a desk or in the great outdoors, all human beings have simple bacteria living on our skin. One set of bacteria belongs to the group called staph organisms. They are pretty tolerant of salt—so they live happily on our skin where many other bacteria die off. And, normally, the staph stay on the outer parts of our skins, coexisting with us without drama.

But, occasionally, a cut or scrape introduces staph into the interior of our bodies, and significant infections result. Hospitals have long struggled with staph infections in patients after surgeries or procedures.

Because staph is a bacterium, it can be combated by antibiotic drugs. But some staph strains are resistant to antibiotics, and a few are resistant to pretty much our whole arsenal of antibiotics.

The abundance of antibiotic-resistant staph on our skin was really driven home to me when I was an instructor in a class at Washington State University. Like doubting Thomas, I had to see to really believe—but see I did!

Other instructors and I had 120 willing freshmen swab their skin and then smear the swab on slightly salty "gelatin" that lay in Petri dishes. We gave each student six tiny paper disks to place on the gelatin. Each of the six disks had been soaked in a different antibiotic, so the gelatin near each disk was influenced by penicillin, amoxicillin, and four other common antibiotics.

Then we covered the dishes and kept them in a nice, warm place for a week.

What the students saw when we examined the dishes again was that most of them had staph strains on their skin that had grown over much of the gelatin, some growing happily right up to several of the little disks. Sometimes staph were even growing on top of some of the antibiotic-rich papers, so able were they to tolerate the drug.

Staph on our innocent freshmen was indeed resistant to many antibiotics. It was quite impressive.

The most drug-resistant sort of staph is MRSA, and it can cause infections that are life-threatening. But my friend Phil Mixter, an immunologist on the faculty at Washington State University, told me about MRSA that has been found not in hospitals—where we might well expect long-term use of antibiotics would select for it—but also out in the natural environment of the world. One study shows that MRSA lives at the beach, up and down the Pacific coast. It's no surprise that mildly salty samples of sand—they're salty like your skin—have staph bacteria in them. But it did take scientists by surprise to learn the sands also have MRSA in them.

"This is natural MRSA," Mixter said to me. "It suggests that antibiotic-resistant staph is not a new thing, and it's everywhere, more than can be due to human overuse of penicillin and similar drugs."

One way to think of how this came to be is that penicillium fungi have been around in the natural world a long time, making "penicillin juice" (as I think of it) that kills microbes near the fungi. Staph bacteria, over equally long periods, have been selected for resistance to the natural antibiotic makers.

There's a microscopic arms race and ongoing war out there, one that can promote a lot of antibiotic resistant bacteria.

So don't get caught in the crossfire at the gym. Wipe down the equipment. Shower with soap and warm water before you return to work.

Let the microbes fight it out among themselves.

Too Much Exercise

I'm not going to give you permission in this piece to live a sedentary life, sitting at a desk at work all day and then on the couch watching television each evening. But medical science increasingly has some evidence of a general principal your mother warned you about: there is such a thing as too much of a good thing.

A few folks really throw themselves headlong into aerobic exercise, running, biking, rowing, or swimming for hours and hours each week. Most of these hard-core endurance athletes start young. Many fall by the wayside in middle age, but there are also those who keep going, completing marathons and similar events well into retirement age.

What happens to the heart muscles of such titans of lifelong exercise? Does the massive amount of work their hearts do to fuel all that labor lead to only good effects—or does the body sometimes register some significant problems owing to all that deep breathing over the decades?

A recent British study set out to address that question. It found men ranging in age from 26 to 67 who were chronic and lifelong exercise kings. The men came from the ranks of Olympic teams in distance running and rowing, and from what the Brits apparently call the 100 Marathon club—a group that requires applicants to have completed at least 100 marathon races.

Okay, we're clear on the picture. These guys were and are fantastically fit. But what could be said about the hearts of the older men in the group compared to the younger athletes and also compared to healthy older men who may have walked around and gardened a bit, but who didn't do crazy things like run 100 marathons?

Researchers used a new type of MRI exam on the three groups of men, the young and old athletes and the "control group" made up of the healthy older men who had never exercised strenuously day after day. The sophisticated MRI was capable of showing truly early stages of scarring (fibrosis) of the heart muscle.

Interestingly, none of the younger athletes had early fibrosis in their hearts—but neither did the older men who didn't follow enormously demanding exercise regimes. Remember, the older men had been screened to be healthy subjects—not guys wheezing at every breath from decades of heavy cigarette smoking, nor guys with heart defects or a history of heart attacks.

But what was striking were the results from the older male athletes. Half of them had hearts that showed some muscle scarring. And the men with the fibrosis were those who had exercised the hardest and the longest.

To repeat, the information from Britain doesn't get us off the hook when it comes to exercise. Let's face it, most Americans are generally in no danger of running 100 marathons over a lifetime.

Again and again, in major medical study after study, the health benefits of exercise have been shown. Aerobic exercise, the sort of thing that makes you breathe deeply, is good for the ol' heart and lungs, and will help you ward off such maladies as Type II diabetes. And if you've noticed lately that you're getting just a bit older, exercise is your single best bet to help combat the battle of the bulge.

Besides all that, let me mention the simple fact that moving around can be fun. An after-supper walk on a summer's evening is a delight and can boost your mood, and a long swim on Sunday afternoon helps at least some of us blow off steam and clear our heads for the workweek to come.

But it's also true that completing more than 100 marathons may be a bit excessive.

Here's one final thought on the British study: inquiring minds reading about such research programs might well ask why women were left out of the research. For years—nay decades—it's been commonplace to leave us females out of many research populations. The logic has been that women are different from men. (Yes, doctors have noticed that—medical school is good for something.) In order to have a simple study with only a few people in it, so the reasoning goes, it's best to eliminate women from the research because their bodies might have differences that would affect the outcomes of the work. Thus, after medical studies of various types, doctors have generalized from results really known only for men but nevertheless applied to women.

Mind you, we could reverse that logic. Researchers interested in heart health, diabetes, colon cancer, or whatever could study only women—because after all including men in such studies could complicate results. Then we could just assume what's true for women simply must be true for men.

That'll be the day.

Sunlight's Darker Side

Each year summer brings us the longest days of the year, with the sun higher in the sky than at any other time. Like many people living in the northern strip of the country, I used to spend a good bit of time worshipping the sun each June. I didn't use any sunscreen because I found that after a few days outdoors, my exposed parts were red-brown—and my chronic joint pains were cut in half.

I knew strong sunlight on Scandinavian skin was a recipe for trouble. But reducing joint pain in-the-here-and-now tended to trump my fears of death somewhere-down-the-road. (I'm rethinking my tactics a bit this year, however, and taking more Vitamin D as tablets rather than making my own from sunlight.)

Sunlight feels like such a blessing—especially to us Northern people—it's easy to wonder how it can kill. The answer lies in the fact that our genetic blueprint is recorded on a tissue of molecules that's both remarkably flexible and exceedingly fragile.

The blueprint—known as the DNA molecule—is a long chain of chemicals wrapped around each other in simple and complex ways. Like a telephone cord that makes tight curls around itself, and looser curls as it flops on your desk, DNA can be wound both tightly on a fine scale and more loosely overall at the same time. It's amazing to me, but the coils are also folded over in complex ways, a bit like little paper cranes from art class. The coils and folds make it possible for an enormous amount of DNA to fit inside the nucleus of your cells.

"It's like packing a great deal of spaghetti into an itty-bitty can," said biochemistry professor Michael Smerdon of Washington State University.

It's mind-bogglingly complex in detail, at least to a simple-hearted geologist like me.

"That's not just you," Smerdon said kindly. "Biology is chaos, a cesspool of complexity."

But one simple truth is that the complex DNA our lives depend on is damaged many, many times each day in many of our cells. That's staggering to think about. I make a lot of mistakes each day at work as a writer, but my DNA creates errors in the basic chemical code of my life at a rate that leaves my many typographical mistakes in the dust.

Our lives are saved (usually) because we have some wonderfully well-developed systems for making repairs to DNA on the fly. Our cells each have great pit crews, if you will, that can put our DNA back to rights more or less every time we complete a lap on the racetrack. The repair crews have to cope with the closely looped and folded parts of the DNA, and the slightly looser bits—no easy task in itself. It doesn't always work out right, but mostly it does, and sometimes the continued mistakes don't matter. A combination of those two factors keeps us going.

But the whole dynamic of frequent errors and endless fixes can change when strong sunlight slices through our skin cells. The energy of the light—the same basic energy that makes sunlight warm—can damage some of the DNA, both in the closely and more loosely coiled sections of the complex spaghetti. Our natural "pit crews" do what they can to help, and mostly they do a good job. But if they miss doing a crucial repair, or the patch is simply wrong, one result can be that the cell gets the message to reproduce endlessly—and that's part of what makes up cancerous lesions. That's why letting your skin burn in the summers is a real risk.

The medical aspect of treatment is not something Smerdon and his research group address, just the fundamental biochemistry that's at square one of understanding the cells' first line of defense against DNA damage. But they know their work can be useful down the road to medicine.

"Einstein said there was about a 30-year lag from research discovery to application of results," Smerdon said. "With biotechnology, it may be less now, but it's still a long time."

Meanwhile, I'll use my Vitamin D pills and try not to burn too badly when sunlight is blessing us with its presence.

Twins, DNA, and Mental Illness

For generations it's been clear that a couple of major mental illnesses seem to occur much more frequently in some families than in others. But it hasn't been so easy to understand why such facts are the case.

One problem is that it's not easy to untangle the dual effects of genes and upbringing, of nature and nurture. Let's focus on one common mental illness about which some facts are known, namely what us old farts call manic depression or what's now termed bipolar disease. Here's one of the fundamental questions scientific research has addressed: are the kids of bipolar parents more likely than typical kids to be ill because they were raised in homes shaped by the disease or because of their genetic inheritance, pure and simple?

One way of starting to address the question is to study identical and fraternal twins. Identical twins share 100 percent of their genes, while fraternal twins share 50 percent. Generally, twins are raised in their households in very similar conditions. For example, they either have an older brother, or they don't—a condition presumably affecting each twin in highly similar ways.

Imagine John and Joe, who are identical twins in a bipolar family, and Ruth and Steve, who are fraternal twins in another bipolar family. If you are a researcher, you can comb through the medical records of lots of cases like John and Joe, as well as many cases like Ruth and Steve. Now let's say you look at all the times one of the identical twins has been diagnosed as bipolar. Next to that name you mark whether his or her identical twin has also been given a bipolar diagnosis. You do the same for the fraternal twins.

Studies of twins have been done in this way both in the United States and Europe. The evidence is clear that the odds of being bipolar if your identical twin has the illness are high indeed. That's not nearly so often the case if you have a fraternal twin who is bipolar. That's good reason to conclude that genetics really matters for manic depressive illness.

Like everything else about complex diseases, the twin studies don't yield a clear-cut, 100 percent relationship. This suggests that developmental and other complicating factors may play a role in bipolar disease. Twins, even identical ones, are a bit different from each other and are treated differently by those around them. For example, they may be split up in the second grade, with John having Mrs. Smith for a teacher while Joe has Mr. Jones. Down the road, small factors in life can make a difference.

And although identical twins do share 100 percent of their genes, science is learning more and more about differences in ways genes are "expressed" and "silenced"—differences that may relate to environment.

Researchers have also studied other kids, namely those who were born into families with bipolar relatives but who were adopted shortly after birth by families without bipolar people in them. Just as with the twin studies, there's imperfect but strong evidence that genetics has a lot to do with your chance of growing up to be bipolar. If your biological parents have the disease, even if you literally never saw them and were raised by non-bipolar parents, you stand a much greater than typical chance of being bipolar yourself.

Nature really matters. So it's good advice to choose your biological parents with as much care as you can!

Today the cutting-edge research in genetics and bipolar disease is focusing on where in our DNA there lies the genes that create the illness. I use the plural term "genes" because it's quite clear that many different locations on our DNA strands are at issue. Some of them may promote the chance of developing a particular form of bipolar disease (there are several flavors of the malady) while others may have to be present in combination to boost the chances an individual will become bipolar.

Researchers in the 20th century moved from simple studies of twins and adoption to technically complex DNA work. This century, we can all hope, will move us substantially closer to useful treatments for bipolar and other major mental illnesses.

Rabid Coyotes and Flu Shots

One December's afternoon a few years ago I crawled through a barbwire fence buried in the snow. It was an experience that led me to relearn a lesson on virology—the science of viruses and their nasty effects on us bigger creatures.

I wasn't graceful getting through the barbwire, so my hand was bleeding from a jagged cut when I staggered to my feet. A moment later my dogs chased a sickly coyote toward me in the corner of the field. Normally a coyote would outrun my canines, but this one couldn't, and all the parties concerned were biting each other thick and fast, surprisingly near to my feet in the snowbank.

Then the coyote started lunging for my dogs' faces, aiming for their eyes.

I was younger then, and with visions of enormous veterinary bills in my head, I stupidly reached down to grab a dog collar. I was rewarded with another slash on my already bleeding hand, I knew not from which animal.

When my canines and I had put some distance between ourselves and the coyote, we declared victory and calmed down a bit. But I didn't know how much of the blood on my hand belonged to me, to my dear dogs, or to the coyote. One thing was sure: the spit on my hand wasn't mine. And, reliving the scene in my memory, I was sure the coyote had been profoundly ill.

Thus it was that, even though it was winter, my doc and I went into close consultation about rabies with public health authorities before the day was through.

Rabies is caused by a virus. If you wait to see if you develop symptoms of rabies before you take action, you'll die. It's that simple. The only effective treatment is to go through a series of shots right away, vaccinations that help give your body's defenses a chance of fighting off the virus. The whole biological battle hinges on antibodies, microscopic parts of us that our bodies can "learn" to create and that bind to the problematic invaders, often neutralizing them and their effects.

Louis Pasteur was the man who first figured out how to effectively inoculate people who had been bitten by rabid wildlife. His work has saved the lives of thousands by giving bite victims a head start on producing antibodies inside ourselves. Luckily, I knew that whole story from the history of science, and believe me I had plenty of time to feel grateful to Mr. Pasteur in the days after that walk in the snowy field.

But even the modern descendants of Pasteur's injections are not easy-as-pie to experience. The first rabies shots entailed waiting around in the doctor's office after the double injection to see if I'd collapse in a coma. Then there was an amazing "rabies headache," too. At home, I ran a fever in the night and, let me tell you, I was bushed.

But, on the good side, the response meant my body was making the antibodies I needed if there was coyote-rabies virus in me.

You likely won't have to think about the rabies virus in your own life. But influenza is caused by a common virus, and you are bound to contract influenza at some point. The regular varieties of influenza kill about 35,000 Americans each year.

Dr. Mary Sanchez Lanier, a virology faculty member, commiserated with me because I'd had the bad luck to go though the whole rabies vaccine process. But she explained that most all of us should get the influenza vaccination each fall. A stiff dose of real influenza can knock you off your feet, leading to high fevers, days in bed, and all the rest of it.

Trust me on this one, compared to the rabies sequence, influenza shots are a piece of cake. And they may keep you alive, at work, and able to take care of your family next winter, when others don't manage those tricks.

Not a bad return on a basic investment.

Dancing with Death
over the Centuries

Once I had a case of influenza so bad I missed close to a month of graduate school. I ran a fever and coughed until it felt like my whole world was turned upside down. Because I'm a geologist, not a medical doctor, I nicknamed that bout of illness "the plague." But what I experienced was a walk in the park compared to the real McCoy.

The sheer virulent power of plague is a tale of human history that's a warning ringing across the centuries. But the story takes its most intellectually interesting turn recently, as science has been unraveling more and more mysteries of the Black Death.

The first widespread outbreak of the plague we know about started in 541 A.D. Called the Justinian Plague because it started under Byzantine emperor Justinian I, it ran off and on for decades.

We don't know much about how widespread the Justinian plague was in Europe because written records were not being kept in many places at that time. But it certainly killed a meaningful fraction of the population we do know about, at one point killing thousands *per day* in Constantinople (Istanbul).

Modern science has taught us that the plague lives in rat populations. It's often transmitted from rat to rat and from rat to people via flea bites. Improved sanitation measures that get rid of rats can help limit the plague. But modern sanitation was not even a distant dream back in Justinian's day.

The story of the plague next jumps like a flea to the Black Death of Medieval times. We know more about it across Europe because more written records were being kept in a variety of places. In London, for example, wise town fathers in 1348 established a special cemetery for those who died of the plague—a cemetery that lay outside the city walls to try to keep the dead from infecting the living. In two years, about a *third* of London's population died of the plague, filling the graveyard quickly, where bodies were interred five deep. European culture of the era was simply never the same after the cataclysm of the Black Death.

The Medieval plague had some different characteristics from the plague of modern years. The Medieval victims sometimes exuded a deathly stench, and the bubonic variety created buboes—painfully enlarged lymph glands. The pneumonic variety was passed directly from person to person by coughing. That mode of transmission made the Medieval plague all the more effective at killing a lot of Europeans in quick order.

Recently more scientific information about the London outbreak of 1348 has come to light. Using advanced technologies of the sort that have been used on remains of animals as old as the Ice Age, researchers have looked for DNA from the plague in the remains of the Medieval dead. The work is made possible by DNA sequencing machines, a high-tech way to isolate, categorize, and sum up the DNA information in small and "broken" samples of ancient DNA.

The research work on the Medieval plague has confirmed that even though there were some differences between the Black Death of the Middle Ages and the plague sometimes found in Africa and India today, the same bacterium caused the malady. The powerful bug is called *Yersinia pestis*. It's not found in European graveyards that date before the Black Death, but it is found in human remains in places like the London mass graveyard established outside the city walls.

The great good news for us today is that antibiotic drugs kill bacteria, the cause of the plague in all its forms. They don't help with viral infections—like the one that laid me low long ago as a student—but they can help enormously with bacterial infection. That—plus improved sanitation—has created modern populations that haven't had enormous trouble with the plague for a long time.

Science is still learning about the Black Death. Let's hope our luck continues while science works out the details of what made the plague so very deadly in the past.

SECTION 6
BIOLOGY

Unlike the other planets in the solar system, Earth teems with life. We like to think we are the highest form of life on the planet, but whether or not that's true we have to admit many of the other species around us are impressive in their own right, too.

Perhaps because living creatures grab our attention, the science of life is always interesting. Biology spans such a wide spectrum—from the structure and function of complex molecules like DNA to the three tiny bones in our ears—there's bound to be something of interest for everyone.

Biology

Visiting Darwin's Grave Ere I Die

My Labrador-mix from the dog pound is quite a mutt, but even so he shows Lab enthusiasm for retrieving sticks I throw into the river. (Just for the record, he's a specialist and won't retrieve sticks thrown on land.)

The best watchdogs I ever had were two Dalmatians. The Dalmatian is an ancient dog, originally bred to loudly sound the alarm if a stranger arrived at the manor in the middle of the night. Dalmatians, as far as I can tell, live by two simple commandments. First: if something happens, a Dalmatian must bark loudly. Second: if Dalmatians are barking, they must *keep on* barking.

The simple process of selecting which of our domestic animals gets to reproduce—choosing which critters get to be the mamas and the papas for the next generation—has long given humans a fantastic array of domestic dogs, horses, sheep, and all the rest. In the 1800s, Charles Darwin had the wit to reason about animals in the wild with our domestic breeds in mind. The core of his idea is termed "natural selection."

Natural selection operates not by fences and leashes that keep potential mamas and papas apart in the barnyard and the backyard, but simply by the processes played out in the wild each day. On average, if an animal has a characteristic that's useful for its environment, it's more likely to survive (and reproduce if it can). If it does all that, then its characteristics are likely to appear in the next generation.

Here's one example of the process.

Several million years ago, there were elephant-like animals all across Asia. We know about them from the fossils they left behind. They were not exactly the same as the modern African and Indian elephant, but if one walked into your living room tonight you'd say, "There's an elephant in the room!"

About two million years ago, global climate changed. The world entered the Ice Age and temperatures became absolutely bitter.

Many animals may well have died due to the massive climate change, while others likely moved south. But some of the northern elephant species changed, as the fossil record shows. Schoolchildren know the animals that resulted as the woolly mammoth and the mastodon. They left many of their bones, tusks, and pieces of fur behind in parts of North America, Asia, and Europe.

It makes sense that woollier coats helped the mammoths survive the massive climate change that had engulfed the Earth. But it can't be that the mammoths simply wished themselves hairier so that they could stay warm in their new environment. What caused them to change?

Darwin saw that natural selection was a key to the puzzle. A woolly mammoth that was a bit woollier than his cousins was more likely to survive the increasingly bitter winters of the Ice Age. Because he survived, he could reproduce, and his offspring would tend to be woollier, too, just like their daddy. Just as Dalmatians can be selected for endless barking, Mother Nature can select animals that are better suited to their colder environments. Enough change, and we've got a new species on our hands.

Natural selection doesn't have to work perfectly all the time to be a profound agent of change. One particular very woolly mammoth may have some bad luck and fall down a crevasse in a glacier, snuffing out his potential impact on later generations. But, on average, if woolliness helps mammoths survive bitter climates, it's likely to become more and more common in the population.

It's unfortunate that our culture got mired down in the bog of political conflict about Darwin's theory. Natural selection surely doesn't seem controversial from where I sit in a church pew on Sundays, and most major Christian denominations accepted Darwin's ideas more than a century ago.

Before the good Lord calls me home, I hope to go to Westminster Abbey in Great Britain. I'd like to pray there, and also pay my respects to Darwin where he lies in a place of honor.

And when I kneel down somewhere in that great building, I'll think of all those long-ago generations of woolly mammoths and the Ice Age in which they lived—and how one clever man came to understand how species on Earth change over time.

Now Hear This

I wasn't quite sure I had heard correctly. My friend, a fellow geologist, and I were standing in the swimming lanes of a lap pool where we had stopped to greet each other.

"My hearing in this ear is a whole lot better," he had said.

Or so I thought.

"The surgery replaced the tiny, tiny third bone—the innermost bone—of the ear," he went on. "It had become ossified, sort of cemented to the rest of my head over the years, so it couldn't vibrate like it should."

From high school biology I vaguely remembered three tiny bones in a little chain in the ear, bones that have the task of amplifying sound waves as they enter the ear. Sound waves in air don't pack nearly the "oomph" as pressure waves in water, so if you want to hear in air (and most of us do), you need little mechanical amplifiers—which is what the three ear bones are. The third and final tiny bone gives the last boost of amplification and also separates the air we live in from the water that fills and conveys sounds within the inner ear.

It's in that fluid that tiny, tiny hairs respond to pressure waves and translate them into electrical signals that flow to our brains. That's the whole goal of an ear, from my point of view as a physical scientist—to translate sound waves into electrical signals.

But the system doesn't work if that last, and most tiny of all, bone cannot flex and move.

"The implanted piston goes in and out in place of that last bone. And it works!" my friend said with evident pleasure. "I can hear sopranos again."

Now, in truth, my own hearing standing in the swimming pool was problematic because I had misplaced my good earplugs that day. My outer ears—the part you can reach with a Q-tip although you are not supposed to do so—had water in them and shaking my head wasn't doing much to get the water out.

All of that got me thinking about air and fluid in different parts of my ear. But only when I talked to my friend Ken Kardong, a biology professor at Washington State University, did I start to understand that my almost random questions about the matter were unearthing a bit of the long history of life on Earth.

Ears are nothing new. Many fish have pretty complicated ears, including fluid-filled inner ears. Fish go back to the Paleozoic Era, the oldest era in the history of life that has complex vertebrate fossils (proper animals with backbones). There was one part of the Paleozoic in which there were many, many fish species in the seas—as we know from the fossil record—but still no complex species at all on land. That's how early and simple was what geologists mean by "the early Paleozoic."

When land-loving vertebrates first show up in the fossil record they are amphibians—animals that move from the water to land and back. Reptiles follow amphibians near the end of the Paleozoic, again a fact we know from the fossil record.

It's no surprise that the inner ears of fish would be filled with fluid. And since the inner ear is separated from the other parts of the ear by a bone and seal, you can see how the inner ears of amphibians and then reptiles would likely remain fluid-filled while the outer parts of the ears started to become air-filled.

When fully land-loving mammals came to dominate the scene in the Mesozoic Era—the era dominated by the dinosaurs—they naturally enough had air-filled outer ears and fluid-filled inner ears. We still do. That's why swimmers need to drain their ears after doing laps so the outer parts are filled once more with air. But we humans also display each day the fluid-filled inner ear that suggests our ancient lineage with earlier animals in the long chain of vertebrates that have lived on Earth for so very long.

May your day be filled with nothing but good sounds—which you can well and truly hear.

Mr. Jefferson and the Ice Age Zoo

Thomas Jefferson had so many serious interests and accomplishments, it's difficult to name even half of them. Besides helping to found a nation, he analyzed the gospels, started a university, promoted fine dining, and bought half our continent from the French. He also squeezed in a few hours now and then to theorize about the origin of some peculiar bones dug out of the earth. (That bit of work made him a cousin to all of us geologists—or so I like to think.) But despite his brilliance in many fields, Jefferson got something important quite wrong, and we perhaps can learn from the fact that so great a mind clearly stumbled.

Up and down the eastern seaboard of North America, some odd sets of bones had been discovered in the gravel bars and riverbanks of colonial America. Many of them appeared to come from an elephant-like creature with enormous, curved tusks. No one had ever seen such an animal in the flesh. Today, every schoolchild would call the remains woolly mammoths and mastodons, but such identifications weren't obvious if you were Joe Farmer in the 1700s and you unearthed just a bone or two from a creek bank.

In 1796 Jefferson seriously studied a different set of fossil bones dug up in western Virginia. The beast was about 8-10 feet long, with a tank-like skull, thick and sturdy bones, and amazingly long claws on its paws. Jefferson first thought the animal was a great cat—a mega-lion of some sort. Later, when he read a scientific paper that came from Europe, Jefferson changed his mind and correctly identified his animal as being a type of sloth. The current scientific name of the beast, *Megalonyx (great-claw) jeffersonii*, honors Jefferson for his connection to the animal.

But what's much more important to me than the old bones is that Jefferson fully expected living mammoths and great sloths to show up in the world. North America was largely unexplored by Europeans, after all, and it seemed quite plausible to him that the animals might turn up, grazing somewhere west of the Mississippi. Jefferson even asked Lewis and Clark to keep an eye out for the beasts when the famous expedition headed toward the Pacific.

To put it simply, Jefferson rejected the idea of extinction. And that remained true until his dying day, many decades later. Extinction simply seemed to Jefferson like the wild imagining of an evil nightmare. If certain animals had dropped off the face of the whole Earth, Jefferson reasoned, others might as well. Then the whole of creation could come apart.

Like many great thinkers of his age, Jefferson could not believe that a divinely ordered Creation would include extinction. It was much easier to imagine mammoths and sloths roaming the Great Plains.

But it wasn't long after Jefferson's day until almost everyone accepted the idea of extinction. In part that was because exploration of the hinterlands didn't yield living mammoths and sloths. And in part it was because more and more exotic animals—what we'd now call dinosaurs, pterosaurs, and ichthyosaurs—were being unearthed, and it was really quite clear to everyone they were not still alive (and thank goodness for that!). And there were more extinct Ice Age mammals discovered, too, ranging from saber tooth tigers to mega-scale beavers.

Two generations after Jefferson, the general public started flocking to newly built natural history museums to see the fossils of ancient monsters on display. The bigger, the scarier, the better, and school children continue that strong current of interest in the subject today in our society, with kids often more able than I am to assign the formal scientific names to dinosaurs and Ice Age mammals alike.

One of the clearest lessons of geologic history is that species are temporary—our own most likely included. We may not like it, but extinction is as common as dirt.

The fact that a man as brilliant as Thomas Jefferson was quite wrong about something so fundamental as extinction should make us lesser minds quite humble. But soldier on we must, with our best judgments and a reasonable tincture of self-doubt to guide us.

Linked by Our Mamas to the Ice Age

You might think you don't know much about the history of Earth's climate. But dramatic climate change is reflected in your genes. Here's the story.

There's a special type of DNA that flows only from mamas to children. It doesn't get shuffled with papas in the process of reproduction like most genes, and that leads to some interesting consequences when it comes to studying human history and climate. The unshuffled DNA that comes from mothers is called mitochondrial DNA. Being a simple soul, I just call it mama-DNA.

Not everybody in the world has the exact same type of mama-DNA. That's because from time to time a small mutation occurs. As a result of those rare events, people around the world have a total of about 25 different types of mama-DNA.

The earliest and oldest mama-DNA groups are from people native to Africa. That genetic fact fits well with archeological evidence that modern people first arose in Africa and spread from there to the Middle East and elsewhere over time.

African natives have what's called L-type mama-DNA. (If geologists had named the groups, we would have used long and obscure names like "Proterozoic" or "Glossopteris" for each one. Mercifully, genetic scientists just used letters of the alphabet for the different groups.)

Two early bands of people that left Africa had M and N mama-DNA that had branched off from the L's. From there, the story gets complicated as many splits from the M's and N's lead to the diversity in mama-DNA we see around the world today.

Native Americans have mama-DNA in groups A, B, C, D, and X. People in northeast Asia have the same types. That fits with the archeological evidence that North America was originally peopled by immigrants who walked from Siberia into Alaska, spreading southward from there.

The passage from Siberia was made easier because the Ice Age was in full swing at the time. A great deal of Earth's water was locked up in glacial ice, so sea level was much lower, making a land passage possible for tribes moving from Siberia into North America. That's just one link between climate history and our genes.

National Geographic is sponsoring a mama-DNA project in which you can participate for a small fee. Through that program I've learned that I, my mother, maternal grandmother, and maternal great-grandmother, etc., are in group H. That fits with genealogical information that shows we go back to American colonial history and, prior to that, to England.

About half of native Europeans are H's. Just like me, Barack Obama may well be an H, because his mother and his maternal grandmother were of European origin.

The H-group arose about 45,000 years ago in the Middle East. The H-clan spread into Europe, but around 15,000 years ago, when glaciers were at their maximum during the most bitter part of the latest period of the Ice Age, we were pushed south into what's now Spain, Italy, and the Balkans. The world's climate then changed once more, and Earth climbed out of the Deep Freeze. We H-folks spread gradually north throughout Europe as the ice sheets retreated. First we percolated up through France and Germany, then up the coast of Norway and northwards in the meadows of Sweden, our genes oozing north as the glaciers melted away.

Ultimately, of course, the H group goes back to the L groups in Africa. Africa is the mother of us all, as both genetics and anthropology show.

But to get back to climate and the fact that some of our genes have been pushed around by glaciers. It's staggering—but ultimately uplifting—to think about the great climate changes that our ancestors faced. Considering they had only the most primitive stone, wood, and bone technologies, they did well to survive, let alone take advantage of the opportunity to walk across land-bridges into new lands. And that makes me quite hopeful about our ultimate response to the next climate shift.

You can learn more about mama-DNA and take an armchair journey with all the peoples of the world back to the Ice Age by looking on the web at National Geographic's "Genographic Project."

Enjoy the trip.

More than One Way to Do It

Those of us who've been around for a few decades have seen friends or family engulfed in bitter custody disputes. There's nothing like denouncing the other parent of your children in open court (and paying steep legal bills for the privilege of doing so). Custody clashes are clearly only the tip of the iceberg of the total cost we sometimes pay for sexual reproduction. From a scientific point of view, there's another problem with sex, and it's a doozie.

Some primitive animals can reproduce either sexually, meaning with a partner, or all by themselves. This means that in some species, females can mate with males, while other females simply reproduce all by their lonesome. Some snails can do exactly that trick. If a female creates baby snails by herself, the offspring are identical clones of the mama snail.

Wait a minute, you say, you've never heard of natural "clones" like that, only special man-made clones like Dolly the sheep.

But actually, you yourself may well have cloned a creature in the plant kingdom, simply by taking a cutting of a plant and letting it root. The new plant is genetically identical to its parent, making it a natural clone.

Now think of a lake full of snails. If a particular female mates with a male snail and produces 100 eggs, she might have 100 babies that are 50 percent male and 50 percent female. Her genes will be in all the babies, along with the papa's genes.

But if a female snail reproduces by herself, she can create 100 babies that are all female and all carry only her genes. Basically, this route puts more of the female's genes into her offspring. Way more!

If evolution is all about surviving and getting your genes into the next generation, it looks like females reproducing on their own should have taken over the whole planet long ago.

After all, it's not so easy to reproduce sexually. You have to find a partner, and that can take some doing (if you are a rare bird, for example). Attracting a mate can be a very big deal, with a big investment of time and energy (think of the peacock). Beyond that, both male and female must do the right things and both must be fertile at the same time.

Compared to that, self-cloning starts to look easy and much more likely to be successful. Why are we not all cloned females reproducing identical copies of ourselves?

Professor Mark Dybdahl of Washington State University is a biologist who investigates that fundamental question.

"There's got to be something in ecology or evolution that favors sexual reproduction," he said to me.

This thought experiment can help. Think for a moment of an Iowa cornfield, planted with a particular strain of corn, all the plants being genetically identical to one another due to inbreeding. Because the plants are the same and genetically programmed to create high yields, the farmer may well get many, many bushels of corn from his field.

But if a disease or parasite arises that can feast on that particular strain of corn, disaster results.

Here's the key concept: sexual reproduction is unlike cloning because it creates offspring that are different from one another. It can be well and good to be a female snail that reproduces exact copies of her own snaily essence. But that's useful from an evolutionary point of view only if nothing comes along that can wipe out all her identical descendants.

"A female snail that reproduces sexually creates offspring that are varied. That increases the chances that some of her offspring can resist problems like parasites or disease," Dybdahl said.

Dybdahl's work has nailed down quite a bit, at least in the realm of snails.

"The snails that reproduce sexually—and pay the costs of having sex—do well only when parasites are beating the clones down," he said.

I don't know about you, but nobody taught me any of this material in 6th grade sex ed class. You have to keep learning well into middle age to really get to the basics—like the fundamental reasons that lie behind clashes in family court.

Tough Way to Make a Living

Wresting a living from the world is never easy. Heck, just making it through a desk-bound workweek is enough to wipe me out most of the time.

But it could be worse.

Imagine sitting throughout your life on a cold rock, making a living as best you can from air and the occasional drop of rain.

That's the task lichens face everyday.

You might say that lichens found a niche that nobody else wanted. Or, as the lichens might say, they have a role on Earth that nothing else has the stamina and fortitude to fill. "The few, the proud, the lichens," might be their motto.

The success of the simple lichen is really a tale of cooperation. Two life forms make a living together in a lichen, each benefiting from the work of the other, like a tiny, socialist utopia.

The first part of the intimate pairing is a green, single-cell organism. I won't be more precise than that, because there are variations in those green cells that go right over the head of a geologist. But the main point is that the green partner harnesses the energy in sunlight and uses carbon dioxide in the air to build carbohydrates.

The green cells would often not do very well on their own. Enter the second part of the pairing, the fungus cells.

Fungi are weird. You know that from eating mushrooms. They are neither fish nor fowl, neither animals nor plants.

The fungi in the lichen combine with the green cells to form a compound organism that's more successful than either half alone. The green cells create the carbs, while the fungus does a couple of different things. The fungal cells make a stable structure for the lichen community and create chemicals that dissolve a bit of the rock the lichen sits on. That, in turn, makes a small amount of mineral nutrients that the fungi shares with the green partner.

The cooperation between cells in the lichen has been a great marriage, one that's lasted for more than half a billion years and shows no signs of heading to divorce.

As a geologist, I drag home a few rocks from around the Northwest every chance I get. When I bring the rocks into my house or office, the lichens on them dry completely. And then years pass.

"But they're still alive, I'd wager," said Professor Jack Rogers of Washington State University. "Just water 'em, set 'em in the sun, and watch."

Rogers is a fungi expert. He was kind enough to come with me on a geology fieldtrip so some students and I could learn more about lichens.

Occasionally, geologists use the size of lichens in our work. The diameters of lichen on a gravestone—a rock surface with a known age conveniently carved right on it—can show you how fast lichen grows in a particular climate. Say lichen on the east side of a tombstone from 1888 are all about an inch across. Using that knowledge, geologists can estimate how long a certain boulder or cliff face has been exposed to the sky, judging by the size of the biggest lichen on east-facing surfaces.

It's elementary, my dear Watson.

There are many different lichens: gray and flakey, light green and spongy—and my own personal favorite, the strongly orange ones. Rogers knows them all and calls them each by their scientific names.

"Lichens are a sensitive indicator of air quality," Rogers reminded me as we talked on a recent fieldtrip. "The Scandinavians have been watching them for centuries."

As an American of Scandinavian lineage, I can testify we are an odd group with peculiar interests. Watching lichens isn't our strangest pursuit.

"The lichens in towns died off when Scandinavians started burning coal," Rogers said.

Raw coal smoke from a simple hearth is a good bit nastier than wood smoke. Some of us geologists actually like the smell of coal smoke on a winter morning, but sane people do not. Our industrial cities used to be pretty sooty places—and those in China still are, in large part because of coal smoke.

"But now the lichens in Scandinavia have grown back because the air quality has improved as folks quit burning coal for everything," Rogers said.

That conjured up a new motto in my mind.

The lichens: the few, the proud, the canaries of our industrial age.

Piecing Together Ancient and Fragile Clues

A younger friend of mine bought a house nearby that was built in the 1880s. The structure had many shortcomings, like the ceiling over the main room that had quite an alarming sag. The new owner threw himself into major remodeling work as soon as he took possession of the place, partly just to prop the old place up.

Digging into the attic my friend unearthed layers of newspapers from the turn of the last century. When I stopped by the house, my generous contribution to all the hard work being done around me to reinforce rafters and crossbeams was to sit in a corner and read the old papers.

Newsprint doesn't age well. Some of the newspapers dug out of the attic have fallen apart into a few pieces, while others have crumbled into hundreds of individual bits. One arresting story I could read fully on some big fragments of paper was about several murders at a farmhouse. It was clear from the story the authorities didn't have a full theory about what had happened at the farmhouse—and perhaps they never understood it.

Modern forensic scientists could go to work on the old tragedy if they had physical evidence like bloodstains preserved from the scene 130 years ago. But the scientists would have to go about some of their work differently than they would on a recent murder case, and therein lies an interesting story.

Just as newspapers can crumble to pieces over time, so can living molecules. DNA is a long string of chemicals, a bit like a chain that's been wrapped around a pole again and again. Over time, links of the chain are likely to fall apart.

Let's say you had a bloodstain from a historic murder scene, or even better from a geologist's point of view, some blood and fur from a beast that lived in the Ice Age. How could you gather genetic information about the material if the long DNA chain is now missing some links?

The answer lies in the fact that DNA is in each and every cell present in the sample. It's a bit like having many, many copies of the exact same issue of an old newspaper. If I had many copies of the same newspaper from a specific date a century ago, I likely could recreate each and every word that was published on that day, just by piecing together different parts of the paper from all the copies available. A particular copy of the paper might be missing the top corner of page A2, but some other copies would have that corner intact. So, using some information from a number of different copies, I could piece together the whole paper.

Using the same logic, scientists have developed ways to use a computer to compare all the broken bits of the DNA chain they analyze from different cells in the same sample. The computer, so to speak, lays out the short pieces of chain, looking for sections that are identical to other sections. By doing many such comparisons, the computer can recreate the whole long sequence, just as I could do with many issues of the same historic newspaper.

Professor Brian Kemp of Washington State University uses this type of approach to investigate matters like the DNA of 10,000-year-old human remains found in Alaska. In previous studies, Kemp has linked the DNA from ancient human remains to those of native people living in areas from California to South America. What we find in both modern and ancient samples are variations on the same basic theme: the DNA of tribal groups here in North America have much in common with people native to Siberia and Mongolia. That fits nicely with the archeological evidence that North America was peopled by folks who came from northeast Asia during the waning stages of the Ice Age.

Investigating prehistoric human migrations through ancient genetic evidence is just one of the avenues now open to us by piecing together many fragments of preserved DNA. Such work was unthinkable just a generation ago, but it's now becoming routine.

Stay tuned for more news from prehistory.

SECTION 7
PHYSICS, CHEMISTRY, MATH, AND ASTRONOMY

All too often physical science and math scare off non-scientists. That's a shame because the beauty and elegance of the "hardest" sciences can be breathtaking. Perhaps recent science fiction books and reruns of television shows like *Star Trek: The Next Generation* help to overcome some of the internal resistance many citizens feel about physical science. In any event, the good news is that each year colleges and universities enroll a fresh crop of freshmen who plan to major in math, physics, or chemistry. Hats off to them—and their instructors!

Physics, Chemistry
Math, and Astronomy

Rapid Chemistry When
It Matters Most

The last time I went to Nevada, I stood on the edge of an enormous open-pit mine at noon. The whistle blew. Then the pit erupted in explosive power, enough to make the Earth rumble.

"I always like to watch it," said the geologist giving me the tour. "It looks like the rocks down there get 'fluffy' when they are blasted apart."

We visited the floor of the great pit, picking up rocks, squinting at them through hand-lenses, and hammered on them. The strong Nevada wind was blowing all day and by the end of it all I was filthy, to say the least.

Although I don't know for sure, if I had gone directly from the mine to an airport, I might not have been able to board a commercial plane. The reason is that some airports employ "sniffers" that can detect explosives, even in trace amounts.

Roughly speaking, there are two kinds of sniffers. First, there are the type with wet noses you can train with dog treats to signal the presence of a wide range of materials, from drugs to explosives to illicit produce, of all things.

Second, there are sniffers that are man-made devices that are also exquisitely good at detecting just a few particles of various materials for which they are designed and calibrated. The second, mechanical type are not as cute as beagles, to be sure, and they don't move around a busy airport lobby on their own four feet, following their noses. But, on the positive side, the man-made sniffers don't get pooped out in mid-afternoon like dogs (and me too).

Unfortunately, no sniffers of either kind appear to have been on duty one Christmas when the "underwear bomber" tried to blow a jetliner out of the skies over our Midwest. If the authorities had used sniffers on Umar Farouk Abdulmutallab, I'm told they would have had a good chance of detecting the explosive in his underwear before he was allowed to board the first plane of his long travels.

I spoke recently about all this with Professor Herbert Hill of Washington State University. For many years now, he has been in the field of ion mobility spectrometry (IMS) (that's the fancy term for a broad set of lab devices that can include the man-made sniffers in airports today). IMS works by giving an electrical charge to small particles, and then watching how they move from one charged plate to another.

Hill is an analytical chemist: somebody who wants to identify chemical samples. He got started in his field partly due to his interest in being able to detect tiny amounts of pesticide in stream water or contamination in drug manufacturing. IMS is his favorite analytic device, his "baby."

Hill's laboratory is room after room of IMS devices, each one set up for a different purpose. While not as interesting as beagles are to me (I'm a dog person—what can I say), the devices are amazing in what they can do. And they have come a long way.

When Hill first worked with IMS, it took about four hours to get a full reading on a single sample. It now takes less than a second—which is why IMS has become so useful in real-world applications like airports. Clearly, in the day-to-day world, we need systems that work fast and that make as few errors, one way or another, as possible.

It's interesting to note that what counts as an "error" is complex. If an IMS sniffer (or a beagle, for that matter) pulls a geologist out of an airplane line just for having been around explosives, that's a "good hit" in the sense that the sniffer is legitimately detecting explosives. But it's not usually useful security information, and is likely to delay not just one rock-head but potentially a whole planeload of people.

As you might guess, Hill's former students work in industry and government. Some of them are likely laboring today to try to keep our airplanes in the skies.

Even though they could pull me aside with their sniffers, I wish them the very best.

Fighting Frozen Toes

If you are outside, a frigid winter's day is often experienced as cold hands and cold feet. Even with good socks and sturdy boots, when I'm outside there are temperatures below which I cannot keep my toes warm (this is more notable the older I get, a trend I don't appreciate).

There are two basic ways to combat frozen toes—and both depend on the chemical miracle of how oxygen (my favorite element) interacts with so much of the world around us.

The first approach to warming up is simply to build a fire. Fire, from a chemist's point of view, is a rapid chemical reaction of oxygen with carbon-rich materials. Where I walk my dog on cold Sunday afternoons in the winter, along the seemingly endless Snake River, fishermen pursuing salmon and steelhead use driftwood to build fires on the shores. With plenty of fuel available, the fires can be big.

If I stop and talk a bit with the fishermen, I can sidle up to the fires—and warm myself at least a small bit. But it's really only my hands I can hold out to the fire's warmth—my feet are still freezing cold.

Enter the miracle of the modern, store-bought toe warmers based on a more controlled chemical reaction between oxygen and iron—with iron functioning like the carbon in wood fires.

Toe warmers are small and thin devices that have adhesive on the back. You open the package they come in, expose them to the air, and stick a pair on your socks above or below your toes. Then you put on your boots, and soak up the blessings of a trickle of heat on your ten little piggies all day long.

How can toe warmers possibly work?

First, it's important to note the bags they come in are fully airtight. When you take them out, they are exposed to oxygen—because the air around us is 20 percent oxygen.

So far, so good.

The active ingredient in the toe warmers is iron. It's not a chunk of iron, obviously, but a fine powder of iron filings, spread out in the toe warmer. The finely divided iron means it's more likely to chemically react and interact with molecules around it—not just sit there in unchanging stubbornness like some middle-aged geologists.

Iron reacts with oxygen in a similar way as carbon, going through the process of oxidation. Iron plus oxygen creates iron oxide, also known as rust, also known as the main material on the underside of my '87 pickup. When iron and oxygen create rust, heat is released. That's the fundamental key to the toe warmers.

There are a variety of other hand-warming devices, too, that run on different principles and can fit in your mittens. But the little pouches of iron filings are my personal favorite because they are so thin they fit in my boots below my toes. In my (ever so slightly stinky) boots, my toe warmers are a small pad that's about 100 degrees F or a little warmer (according to my measurements). That's blessedly cozy for my 10 toes on a cold afternoon.

The iron-filing toe warmers are one-use-only. That's a shame, but that's the way it is because there's no chemically easy way to pull the oxygen atoms off the iron once the two have combined. So you toss the toe warmers into the trash when their heat dies down to nil.

But, on the bright side, you and your clothes don't smell like a bonfire for the rest of the day.

Fear and Four Lab Explosions

On a couple of occasions in graduate school I stupidly miscalculated the effects of mixing strong acid and water—and then adding heat. There's nothing like exploding acid droplets quite near your face to give you pause.

The second time I managed to make the same, simple error I walked home, a journey of about eight miles. As I strolled, I gave serious consideration to going into the law. But after that long walk, I realized I didn't want to make a decision about the direction of my education based solely on fear. So I stuck with the sciences.

But my tale of lab explosions doesn't seem like much compared to two others I know, and they inspire greater fears in many members of the general public.

I work every day with a chemist here at Washington State University who likes to say she is so radioactive she glows in the dark. That's not true, of course, but she did get a large dose of the highly radioactive chemical element called americium, which is a byproduct of plutonium. The tale goes back to when she worked 25 years ago at the Hanford nuclear site in Washington State.

Hanford is the place where America made plutonium during World War II for nuclear weapons. After that, it became a major installation for our production of nuclear weapons in the Cold War. I know some people who are afraid even of the universal radiation symbol, but at Hanford it's almost as common on old signs as the tumbleweeds piling up in the wind along the fences.

That day at work 25 years back, my colleague knew she was in trouble when the reaction container she and her boss were working with at high temperatures and pressure started to leak—spewing hot water with radioactive materials on her arms, her lab coat, and her shirt.

"But my boss and the folks in environmental health and safety were calm," she says, "and so was I. There were emergency showers, and of course I gave up my clothes, and then they scrubbed a layer of skin off my arms."

Harold McCluskey was another Hanford worker who received a much greater exposure to radiological materials. A chemical reaction with the material he was working with showered him with americium—about 500 times the occupational standard for exposure levels.

McCluskey likely would have died but a doctor gave him an experimental drug that removed about 80 percent of the contamination. McCluskey had to remain in isolation—and I do mean isolation—behind concrete and steel until the treatment ran its course, but he then returned home. He was known as the Atomic Man in the Northwest. He died in his mid-70s, about 11 years after the accident that made him famous. He had never regained his full strength after the exposure, but he died of heart problems, not cancer.

Everything about plutonium and americium stirs up strong emotions. That's understandable. But not everyone is aware that technical people have been exposed to radiation in moderate to very high doses—and lived to tell the tale. Every day I'm reminded by the presence and hard work of my colleague that americium is not, in fact, necessarily the end of everything—or anything. It's a real and serious threat—as are other industrial dangers—but it's not the only thing abroad in the land that can lead to an early demise.

At some point in this century Americans will have to decide if we want to seriously embrace nuclear energy or not. Nuclear power offers the hope of energy independence, and it can meaningfully address many greenhouse concerns. But, for some citizens, safety and waste disposal issues outweigh those advantages.

What I hope is that the discussion about the nuclear issue can be made on the grounds of facts and rational discussion—and not just fears of chemical exposures. Like that day when I walked eight miles to calm down, we need to be thoughtful as we make our decisions.

Eating Our Experiments

My 85-year-old mother bent over the cookbook and read aloud to me as I wolfed down a chicken sandwich I'd made at lunchtime. The reading was a lesson in how to make a traditional—and very fine as it turned out—pork roast.

Personally, I suspect it would be morally responsible to live as a vegetarian, and certainly good for my family's health and for the nation's medical bills. But I'm a sinner, and my kitchen produces meat and poultry meals on a daily basis.

My mother read aloud the part of the recipe we both knew well concerning what happens when you take a beef or pork roast out of the oven. For a good bit of time, the meat will continue to cook as it rests on your countertop. And, indeed, the temperature inside a roast actually rises for about 5–10 minutes after you take it out of the oven.

What's up with that?

While it's cooking in the oven, the roast experiences a moving "wave" of intense heat that's coming from the oven into the roast from all of the meat's surfaces. At first, only the outermost smidgen of the roast is warmed. Then it becomes hot and a little bit more inside the roast becomes warmed—and so on. Over a couple of hours, the wave of heat that started on the surface of the roast has penetrated into it, further and further.

When you take the roast out of the oven, that wave of moving heat starts to collapse. But the decline takes a bit of time, and the wave is still moving inward. So the inner portion of the roast can and does warm further.

Geologists love heat waves on a much bigger and slower scale. Here's why:

Last summer, the solid rock and soil around your home was warmer than it is now. That wave of warmth went down into the Earth all summer long, growing like the heat wave in the roast in the oven.

The wave that's down under our feet got launched going into the Earth, and it will continue to move downward. By mid-winter, it will be about 60 feet below the surface. Yes: last July is really down there, about six times the depth of my basement's floor.

And the heat wave from the previous summer is down in the Earth, about 120 feet deep.

I like that idea very much. I'm not sure why, but the history of old heat waves is pretty cool.

Physics, Chemistry, Math, and Astronomy   133

Yet it's also true the waves are getting a whole lot smaller in size. Really smaller, and rapidly so.

This winter, last summer's heat wave will be only 0.002 as big as it was at its peak. And the summer warmth of the previous year will be reduced to 4 parts in a million of what it was.

The numbers just mentioned came to me from Fred Gittes, a physics faculty member at Washington State University. Because he's not a rock-head, but a clear-thinking physicist, he worked out the mathematical details as formulas and graphs for my pork roast, just for fun. But he had to use basic figures for the conduction of heat and the like, assuming that the pork roast would behave as water does. That's because physicists don't have a completely detailed picture of just how heat gets into the meat in your oven. That's clearly our loss.

The roast that inspired this train of thought is long since gone. The next roast in my house is coming up shortly, but it's going to be a fancy "crown" roast with a big, gaping hole in the middle. So it's quite a different kettle of fish, to mix images of different repasts.

Physics is fun, in particular, Fred and I can agree, when it's applied to food. Eating experiments has a lot to be said for it.

It's Your Beautiful Eyes

Although Kevin R. Vixie is a mathematician who works just two doors down the hallway from me, his work is a world away from mine. I do easy stuff with rocks and words. Professor Vixie spends his days immersed in truly complicated equations.

One part of the abstract and difficult mathematics in which Vixie is immersed each day can help computers recognize faces—such as those of people passing through the security areas of an airport. That practical application helps me feel I have a hold on what he is doing with his otherwise impenetrable equations.

The path Vixie and colleagues are following in their branch of mathematics is just the current twist of the journey that's taken him from some tough times to a very interesting point in his professional life at Washington State University and beyond. And his life gives me hope for students and creative people everywhere, including those who don't necessarily fit in with the crowd too easily.

Vixie was homeschooled from the eighth grade onward. His family lived in rural New Mexico at the time, and he was free to "wander in the desert and become quite obsessive about playing the violin," as he puts it.

That background, he thinks, helped him stay focused and free to learn. Most 8-year-olds, both he and I agree, are naturally full of that internal freedom. All too many 18-year-olds, for whatever reason, have lost touch with it.

But life deals some very tough knocks to even the young and gifted. The death of both of Vixie's parents while he was young was a blow that took years to absorb. After college, Vixie spent considerable time in a diversity of pursuits.

"I did the living-in-a-cabin-in-the-woods thing in Oregon, and taught grade school for a year, and ran a lab in a medical school," he says with a laugh.

Eventually, Vixie returned to graduate-level mathematics and blossomed there. That path led him to Los Alamos National Laboratory for intense study. And while there, he started to focus on the branch of mathematics that makes image analysis possible.

Vixie put up some of the basic equations from his field on the chalkboard in my office recently. (I'm pleased to report I have a real chalkboard in my office, one with chalk—a natural rock—available for writing whenever the mood strikes. But I digress.)

Vixie's equations don't seem nearly so clear to me today as when he was helpfully explaining them bit by bit. But here's one humble way I've thought of explaining a splinter of how abstract mathematics can help with practical matters like face recognition.

Back in high school, your math teachers had you graph lines and sweat a bit over the equations that describe them. (The field of algebra can fully describe a straight line. Score one for the Arabs who invented it.)

To a mathematician, the image of a face cannot be fully described by equations. But it can be approximated by them. And approximations can be good enough. (Score one for Vixie and his colleagues.)

The name of the game in face recognition is to improve the approximations. The face recognition business depends on both abstract math and on computer programming. Both parts of the biz are getting better over time. With enough computing power and a high-quality camera, faces of folks at security gates can now be meaningfully compared to digital photos stored in data banks.

The choice to do so or not, of course, is up to us as citizens in a democracy, Vixie said.

"But even if we don't choose to do more surveillance," Vixie said, "there are other applications for the same types of mathematics."

Not many people can do highly technical work like Vixie. But he is adamant we all can do something well. He found a new start in mathematics at an older age than most, and he believes bright students who may struggle for a time can come roaring back—even in what many of us consider difficult arenas.

Vixie's outlook renews my hope for kids soon to plunge into the new school year—and for us old biddies, too.

Real Life Star Trek

I'm old enough to remember early reruns of the very first Star Trek programs. But it was the second Star Trek series, *The Next Generation*, that many of us remember most clearly.

TNG (as fans call it) was masterful at blending real physics and fiction. Once, Stephen Hawking, the great British physicist, made a cameo appearance. He has been crippled by Lou Gehrig's disease, and sits in a special wheelchair. Essentially, he cannot move nor speak, but he communicates via computer and a synthesized voice. He is an active scientist—and a Trek fan.

Besides the real-life Stephen Hawking, TNG introduced millions of people to some fundamental physics, including the concepts of black holes, quantum fluctuations, event horizons, and more.

TNG was hardly the end of Trekking. Other television series and movies were made based on the core idea of Star Trek, until the whole great mission finally petered out in deep space (to the great relief of some innocent parties married to Trekkies, I'm sure).

Naturally, the real parts of Star Trek—the fundamental truths of physics—are both alive and well. Here's how I think of them: our universe is made of everyday matter and forces we can directly see, but it's also made of invisible or larger scale things that only modern physics and mathematics can describe and help us understand.

When I drop my coffee cup off the edge of my desk (which happens all too often, I'm afraid), it falls to the floor. Clearly, gravity is at work. Beyond that, the cup exists in three-dimensional space, and the whole falling process occurred in time.

But, as Einstein knew, gravity is much more than a simple force tugging at coffee cups. Here's an analogy that may help a little with the core idea—but be patient and work with me on this one, because it will be strange at first.

Imagine that the space and time around me and the coffee cup are a bit like fabric—fabric, you could say, with threads in one direction representing space, and threads the other direction representing time. The gravity made by the Earth can be pictured as a depression in this fabric. So, when the coffee cup falls, it moves downhill into the depression that's in the "fabric of space-time."

The really cool thing—as TNG fans know well—is that there can be ripples in space-time fabric. These are called gravitational waves, and they cross the vast empty space of the whole universe. On Star Trek, the gravitational waves were represented by shuddering space ships. In the realm of science, they are described by the elegant mathematics of Einstein's general theory of relativity.

Some gravitational waves are the left-over echoes of the Big Bang, the event that started it all. Those are the waves of most interest to scientists. There's a facility near where I live in Washington State that's built on a large scale to detect the waves. Imagine two arms of a very special pipe, laid out on the ground at right angles to each other, each more than two miles long. Inside the pipe are lasers and mirrors to detect the slightest distortion of length—distortions based on gravitational waves passing through them.

Professor Sukanta Bose of the physics department at Washington State University is one of many hundreds of physicists around the world who collaborate on all the complex work needed to make sense of the signals they get from detectors such as that one here in Washington and a sister detector in Louisiana.

Because although Einstein's general theory of relativity predicts them, and tiny movements in large objects confirm them, scientists have still never actually *seen* gravitational waves.

"I want to see them in the flesh," Bose said to me recently. "Until we detect them, they are a belief, not yet turned into truth."

Bose and others who work in the realm of fundamental physics reflect the core of the scientific spirit. It's abstract work, but I'm hoping to report someday soon that gravitational waves crossing space-time have been detected by Bose and his colleagues around the world.

It'll be a great day for science, and for Trekkies too.

"Make it so!"

Complex Times and Simple Lines

At sunset sometimes in late September or March, I like to confirm what I know of Mother Earth in my backyard with a couple of stakes and twine. With just those simple tools, you too can see some of the fundamentals of the Earth and the solar system, as well as part of how humans can tell time around the planet.

If it's clear weather as the sun is going down on any night around the 21st, you'll be in business. Just pound a stake into the Earth. The shadow from the stake will be running due east no matter at what latitude your town may be. Then pound Stake #2 in the middle of the shadow of Stake #1. Next, connect the two stakes with a bit of twine. You now have an east-west line segment, tied off on each stake.

I performed this ritual in the front yard at my previous house—just to double check it had really been laid out with respect to geographic compass points—and I was rewarded with the knowledge that it was. But I've moved since then, and I want to do the same task in the yard of my "new" house, just to make sure it was laid out correctly back in 1949 when its foundations were poured. (I'll sleep better knowing my toes are pointed toward due geographic north in my bedroom. That's just me.)

Once you have the east-west line segment made of twine, you can use a carpenter's triangle to place a north-south line at right angles to your east-west line. Then you'll have the four geographic compass points represented on the ground: geographic north, south, east, and west. (Although it amazes many people, magnetic compass points are entirely different from geographic ones, but that will have to be the subject of another column.)

If you preserve the east-west line segment you made on the Earth through the months to come, you'll have a way of telling seasonal time with respect to a physical line on the Earth. From late September until the spring equinox in March, the sun will both rise and set to the south of your east-west line.

The absolute nadir of the whole experience will fall around December 21, the shortest day of the year in the northern hemisphere. But, by the spring equinox, the sun will have "moved" gradually back on the horizon so that it rises and sets along your geographic east-west line. Then, in the sunny part of the year, sunset and sunrise will occur north of your east-west line and the days will be wonderfully long.

But there are several interesting problems that lie at the heart of telling time by the sun—and also knowing where we are on the Earth by using sunlight as our guide. Sunrise, after all, occurs minute by minute, moving westward across the planet, in a continuous process that goes on and on. But even though sunlight continuously blesses the Earth in this changing way, we humans want to divide Tuesday from Wednesday, and separate 2:49 from 2:50, all with what lawyers call "bright lines."

Bright lines superimposed on natural and continuous processes involve compromises. The Greenwich Meridian Line in Britain and the International Date Line is one enormous imaginary line that circles the whole Earth. It's a kind of compromise, although a highly useful one. When you cross the line in the Pacific, the day of the week rather magically changes. The other part of the same line in Greenwich, Britain, is home to Greenwich Mean Time (GMT), which is the time you'll hear announced on BBC radio.

Greenwich is also home to the Prime Meridian, the "zero" of the system of longitude we agreed to use long ago. Here in the United States, we live to the west of Greenwich. That's why our degrees of longitude are all demarked "west" on maps. In Russia, longitude lines are "east." Greenwich was an arbitrary place to plunk down the central or zero line of longitude, but it works.

If you want to help yourself (or a young person you hold dear) understand the points of a compass, as well as lines of longitude and telling time around the Earth, try playing with stakes and twine in late September or late March. Earth herself is revealed to us by such simple things as shadows in your own backyard.

Secret Codes All Around Us

One of the better phases of childhood, it has always seemed to me, is playing with codes and secret messages. You may remember a summer's afternoon with "invisible ink" made from lemon juice. Perhaps your playmates devised code games for writing based on substituting numbers for letters, or you spent a day slowly beating out an important message in Morse code for the neighborhood kids to hear.

Despite what you think, you may not have left the world of intrigue and secrecy behind. If you buy anything via the Internet, you're using sophisticated codes because all secure websites are based on them.

In childhood, the simple, number-for-letter code worked the same basic way whether you were encoding or decoding a message. That's a "one key" approach that's still used—in much more sophisticated form—in some modern cryptography. The encryption method used by governments has this kind of structure, although it codes information based on groups of letters rather than one letter at a time. In the common method, this means that instead of 26 options to write down—the 26 letters of the alphabet—there are oodles and oodles of options (actually, a little more than the number 1 followed by 77 zeros).

The size of that digit alone tells us we've entered the realm of mathematics, with computers necessarily to do the work of calculating and keeping track of everything.

The trouble with the one-key approach is that you have to get the key to everyone who needs it without nefarious folks learning the key as it moves around. And security is likely to go downhill over time in part because multiple parties have the same key and use it on an ongoing basis.

An alternative is the "public key/private key" approach in which the key to encoding a message is known to everybody, but for practical purposes an entirely different key—called the private key—is needed to decode the message.

Here's the way mathematics professor Anna Johnston at Washington State University explained it to me.

"Public key/private key cryptography is based on special mathematical problems. These problems are next to impossible to solve in one direction, but fairly easy in the other. Think of the hard direction as climbing Mount

Rainier on your knees, with the easy problem more like taking a helicopter trip down from the summit," she said.

Let's say you are trying to communicate securely from your computer to Billy Bob's computer in Baltimore. Starting with a large, random prime number to build on, your computer can make a public key for your coding and a private key for your decoding. You can send your public key to Billy Bob, but he (and any Internet hackers or spies) would have to "climb Mount Rainier" to find your private key.

Only your private key can decode Billy Bob's messages to you. His first message to you can contain his public key, so you can code your reply to him using his public key—but only his private key can decode what you just sent.

Even throwing huge amounts of computing power at breaking the public key/private key system doesn't work.

"But advances in mathematics and computer science need to be watched, in case something comes along to make the current systems insecure," Johnston said.

You've doubtless noticed that, from time to time, a bank or even the CIA has lost private information from secure sites to hackers. That's partly because the trouble with even the sophisticated public/private key is that the security of the private key remains crucial.

Code-breakers need only learn Billy Bob's private key (or yours) to snoop. A computer virus that reports individual key strokes or otherwise reads the secret decoding key can give hackers what they want, and all security is gone in an instant.

By the way, Johnston is a mom as well as a mathematician, and sometimes she teaches secret codes to kids.

"They love codes until I mention the 'M-word'" (meaning mathematics), she said with a laugh.

But talking with Professor Johnston, even this aging rock-head feels a bit more clear about modern codes.

SECTION 8
SCIENCE HISTORY AND EDUCATION

T he way science has changed over the centuries and the
theory of evolution are both subjects dear to the Rock
Doc's heart. Although some scientists are reluctant to
admit it, science is a human endeavor, not a discipline that
impersonal robots can perform. And because it's a human under-
taking, it has at times been associated with some unfortunate
turns (think of eugenics).

Another place where science and politics have intersected is
shown in our attitudes about the theory of evolution. Our society
has been marred by what I think is unnecessary discord about
evolutionary theory. In my opinion, evolution is no more a threat
to religious life than is the theory of plate tectonics or the periodic
table of the elements. And I'm pleased to report that I've taught
basic evolutionary theory in freshman geology courses where, in
general, students have been glad to learn about something all too
often skipped over in high school biology classes.

Science History and Education

Teaching Compromise Rather than Science

The public is bombarded with news reports saying that young people in the United States aren't learning enough about science, especially compared to kids in Asia. I'm not sure that's true, because I work at a large university where I see very able American students starting to excel in their scientific careers, and I hear back from them as they flourish in later years.

But perhaps we really are falling behind. After all, nearly everyone says so. How could we start to investigate whether that's really the case?

It's tough to imagine a single, truly excellent science exam that we could give to all kids in places as varied as rural China, urban Germany, and small town USA. Apart from the problems of translation and grading, there's no reason for the schools in those different places to cooperate with such a test. It's not impossible to directly measure student achievement around the world, but in practical terms it is a challenge.

Perhaps we better just focus on kids in the United States and think about what their science education is like, especially in the crucial grades of high school. What are the meat and potatoes in science classes we are delivering to most of our young people?

One fairly easy way to discover what is being taught to American kids about science is to ask their teachers. Biology teachers are of special significance in this regard, because many kids in high school take biology but no other science classes. (That's an amazing statement in itself—but that particular fact will have to be the subject of another column.)

I'm sad to report what most high school biology teachers in the nation's public schools say they teach. According to an article published in the prestigious journal *Science*, only 28 percent of biology teachers follow the recommendations of the National Research Council and teach the basic theory of evolution by natural selection.

As I do the arithmetic, that means an astounding 72 percent of our high school teachers don't teach the organizing principle that stands at the base of modern biological science.

It's not that the National Research Council recommendations are really so difficult to achieve. The idea behind them is just to present the evidence behind the theory. There's a lot of evidence in favor of evolution, ranging from the fossil record that shows more complex plants and animals appearing over time to genetic similarities in groups of organisms.

Humans can "make" Chihuahuas and Great Danes out of basic dog stock by choosing which individuals get to reproduce, thus shaping the next generation of canines over time until we get what we want. We geologists have a lot of time at our disposal—because Earth history is long—so it's easy for me to see that Mother Nature can also select mamas and papas over time in a way that leads whole species themselves to change.

What can explain why 72 percent of our nation's high school biology teachers don't teach basic biological theory in class? The problem isn't limited to one region of the country, so it's not just a Bible-belt issue.

I suspect the personal religious convictions of the silent 72 percent don't explain everything that's happening. I go to church every Sunday; I've been doing so all my life. (Heck, on a good week, I even put money in the plate.) So I can testify that most of the traditional, mainline Christian denominations in this country accepted evolutionary theory around the year 1900.

True, some high school biology teachers really are Creationists, rejecting every form of evolution. Their views, of course, are not scientific. People are welcome to religious ideas—very welcome in my book. But beliefs given to us by our faith are not what should shape science education.

I suspect a lot of teachers in the silent 72 percent are simply taking the easy road. They know that if they teach evolution, some parents will be upset. If the teachers just neglect to mention the theory, who is the worse for it?

But it's our kids who suffer from the fact that we are not teaching the basic theory of biology. We're not teaching them good science—and by our example we are teaching them cowardice.

Perhaps we really are falling behind, both in the realms of science and, even more importantly, of integrity.

The Threat to Recovery
You've Not Heard About

In recent years we've all seen many true advances in medical science, energy technologies, and more—the kind of win-win work we need now more than ever to find new ways to build our future and propel us out of this global recession.

But scientists come in two flavors, what I call the workhorses and the racehorses. There are good reasons the Amish don't plow fields with racehorses, and nobody wins the Kentucky Derby riding a workhorse. America needs both kinds of scientists. The trouble is, we are starting to be in danger of losing one set, and the general public knows nothing about it.

Here's the good news. The workhorses of science and engineering are the folks who have a bachelor's or master's degree and a lot of enthusiasm for the workaday world. America is producing well-trained scientists and engineers who fit this description—that's why exceedingly smart and dedicated young kids from around the world still come to the United States to go to college here in science and engineering. Students from our own country—very much including those educated in public schools—also hold their own in the undergraduate classroom and have rich potential in technical work once they earn their degrees. You can see lots of both sets of the young workhorses by visiting any college campus in the country at every point of the school year.

Another bit of good news is that universities now do much better educational work than they commonly did in the past. Here's what I mean: there's no better way to learn how to discover new technical knowledge than to jump in and do so, not just sit in a classroom and hear about the work. At both large public universities and elite private colleges, that kind of learn-by-doing work is now encouraged and supported for undergraduates, and it will yield dividends throughout entire decades yet to come in the professional lives of those who are truly being taught how to think.

These are, however, tough times. Many public universities are cutting budgets because revenue streams from state governments are shrinking substantially. Private universities have also been hit hard because endowments have dropped in value with the stock market. But this isn't 1931, and education in science and engineering in the United States is not in free fall, thank goodness.

It's the racehorses of science who look like they are in real trouble. These are the scientists who earn Ph.D.s and normally stay in their disciplines for 45 years. They spend their intellectual lives making many of the most fundamental advances in our knowledge base. Some of them direct large and well-staffed laboratories, others work by themselves using not much more than mathematics. However they go about their work, they are crucial to the recipe that makes for some of the greatest forward progress we've seen—advancement we very much need to keep alive in this recession for all sorts of reasons.

Many students earning Ph.D.s in the sciences each spring simply don't have jobs to which they can apply. That's not entirely a new problem, but it's a trend that's getting rapidly worse. More and more of these students do longer and longer "postdoctoral" studies, work that essentially extends their time in university labs while they wait for jobs to appear. And wait, and wait, and wait some more.

The rapidly increasing number of postdocs is worrisome. We've invested a whole lot in them. If we can employ these very able and highly educated young scientists—many of whom are not native to the United States but who would like to remain here if they can—we'll be better off. If we never fully employ the postdocs or they go elsewhere in the world, we will be vastly the poorer for it. The "brain drain" that could run from here to foreign parts is staggering to contemplate.

How to quickly and fully employ America's postdocs is a national problem, one I hope will spark a public discussion.

Keeping It Simple

A t every level, humans have a natural drive to understand the world around us. We make a lot of progress with that task in some areas—like basic chemistry. But we don't make much progress in some others—like the "chemistry" that springs up between two human hearts.

But *why* it is that science makes progress on many fronts is puzzling. Why should we be accomplished at creating things like the periodic table, the theory of relativity, or Maxwell's equations—human constructs that are enormously helpful in understanding the world?

At the level of recognizing a tiger and knowing to run away from one, it's no surprise why we're good at understanding Mother Nature. Evolution would weed out those who have trouble grasping the predator-prey relationship. But, at the same time, there's no clear evolutionary reason we can see that people who are good with very abstract reasoning (like Einstein) would spring up and do so very well at their labors. To put it another way, why can we calculate the mass of an electron or do a thousand and one other tasks that are routine in research science and engineering?

Even if basic problem-solving is a hallmark of modern *Homo sapiens*, we are remarkably good at it at a level that's astounding—yet we don't know why we are suited to abstractions.

And the conundrum gets deeper the more we think about it.

The methods of science are a hodgepodge we inherited from the intellectual past of ancient Greece, the Middle Ages, and the Renaissance.

Here's one example. Einstein's most famous equation is so simple it can be printed on a t-shirt. (That's my idea of good science—t-shirt printing!) Indeed, everything studied in freshman college physics—all the powerful equations that matter—can be printed on the back and front of a t-shirt.

We scientists are taught that simple is good, that a simple explanation of what looks complex is worth our serious consideration. To put it another way, if we have two competing hypotheses for something we've studied, and one is more simple than the other, we are (all things being equal) supposed to prefer the simpler one. This is called "the principle of parsimony" or the "principle of simplicity." As students like to say, "keep it simple, stupid," a phrase that is captured in the KISS principle. But it's also called by the rather difficult name of "Ockham's razor," and therein hangs a tale.

Keeping explanations simple sounds like common sense. And, I suppose, it is. But the rule is actually one we inherited from medieval times. A great intellectual of that era, a guy named William of Ockham, was in quite an argument with his colleagues about several things. The great discussions of that day were all about God, so the great disagreements were too. Our friend Will argued that theologians should prefer the most simple ideas or explanations they had about theology because simplicity has a note of elegance and power to it—just like God. (Will got that notion mostly from the Greek tradition, I've been told.)

In the Renaissance, when people started to more seriously study what they assumed was God's creation in the natural world around us, it made sense to import Will's principle straight into early science. Hence the name—Ockham's razor, which gives credit to him for the idea of cutting through complexity to the elegant and powerful simplicity likely to offer the most explanation.

But what's interesting is that modern science—which no longer expects to see God's handiwork in creation—still uses Will's idea to good effect. And, of course, it's doubly interesting that other parts of our creative lives, like good literature, poetry, and portions of our spirituality, also seem driven by the quest for the profound and sublime that often comes with simplicity.

It's just not clear why Will's "razor" is still so useful in science and elsewhere, nor why our minds can deal with the abstractions of science so successfully to start with. Maybe it's all chance, a by-product of our problem-solving skills developed many millennia ago to make better and better stone tools in the most efficient way possible.

But it's all surely good fortune. For science, engineering, medical advances, and the rest depend on our intellectual heritage going back to the ancient world and to Medieval Europe, as much as they do to our current creativity.

Knowing Nothin' Can Be Good

I wasn't expecting a brochure with a definition of science in the waiting room of my doctor's office. But there it was.

"The aim of science is not to open the door to infinite wisdom, but to set a limit to infinite error," the leaflet stated in a bold quotation.

The rest of the pamphlet concerned an explanation of the basic effects of cocaine on the brain. But I cogitated only on the quote while waiting for the doctor. When I finally got in to see the M.D., I showed it to him and asked his opinion. (It's probably not so easy to be my doctor.)

One way I have thought in recent weeks about that rather difficult and dense quotation is bound up in my answer to a question a friend asked me recently via email.

"What's your correct mailing address?" he wrote.

I tapped in my reply.

"It's surely easier to type out the one correct address than the infinite incorrect ones," I wrote. (It's probably not so easy to be one of my friends.)

In the realm of research science and engineering, people work like dogs each day to expand our knowledge of the natural world. At least, that's one way of looking at it.

But with each thing we learn, we learn many things that are not the case—the false addresses, if you will. And, in truth, what we often unearth in research science is a false address, followed by another, followed by another, followed by another. It can be a very long road of false addresses, walking up and down streets that are even entirely in the wrong city, until we get to something like a good, useful, and true piece of knowledge about the natural world.

Understandably, that's a frustration. It frustrates people who are ill and are waiting for a cure for their disease from medical research—and instead get promises of research in progress that will take years or decades to come to fruition.

It frustrates people with environmental concerns who wait for one clear scientific theory about a particular problem—but who learn of a series of changing hypotheses instead.

And it frustrates scientists and engineers, too, at least much of the time. (We're people too. Even if we don't always sound like it.)

But my strong belief is that American research in science and engineering has been a major ingredient in transforming the nation—and the world—for the better in recent generations. Diseases that killed and crippled children in my parent's day are all but insignificant in the United States today.

We have knowledge and technologies for addressing environmental contamination. We cannot predict undersea earthquakes, but we have technologies in place to detect the tsunamis they sometimes produce and warn as many people as we can.

Compared to the "good ol' days" we have engineering marvels in everything from communications to reliable and energy-efficient cars. (Yes, energy-efficient. Despite all the complaints about the American car fleet, it's wonderful in all respects compared to the cars I grew up with in the 1970s.) We are linked by global communications in the Internet that are virtually instantaneous. We are beginning to have alternative energy generators that hold real promise for larger scale applications down the road.

A large measure of all that progress has been made by limiting our errors, step by step.

We all know the story of the zillion of attempts Thomas Edison and his assistants made to find the filament of a light bulb that would both burn brightly and not burn out. Edison walked along a lot of streets, discovering false addresses that would not work. But each step he took was still good work—it just wasn't work with a positive result.

Patience is a virtue, in science just as in life. And it's probably useful to keep in mind that "infinite wisdom" will always be outside our grasp in both realms. Pruning back ignorance and falsehood, the wrong addresses in useless side streets, is a valuable function of science, just as it is in ordinary life.

That's good news—it's the reason why Edison and colleagues were successful in the end.

New Eyes and New Ideas

Recently we passed the 400th birthday of science and engineering. That mile-marker is worth noting and giving thanks for because each day those twin disciplines improve the lives of billions of people around the world. (Beyond that, science and engineering are awfully fun, so their total effect is sort of like combining doing good all around the planet with the pure joys of playing chess. Or so it seems to me.)

The game all got started in 1609 when a number of people in different parts of Europe started to put two glass lenses together in a new way. One lens was concave (depression going into the lens), one convex (the surface domed outward from the lens). If you join the two by a leather tube, you have a simple telescope—a device made by at least 1609 in the Netherlands and likely in several other countries around the same time.

Galileo Galilei jumped into the telescope game with both feet as soon as the new device turned up—which doubtless contributes to the false but common notion that he invented it entirely on his own. He didn't invent the telescope, but he did most certainly see the important applications to which the best telescope of the day could be turned—a device about as powerful as an inexpensive telescope you might give a grade-schooler for Christmas today. With that as a starting place, Galileo went to work making night observations of the moon, the stars, and the planets. His progress was rapid, and the implications of his findings launched the scientific revolution and the modern era—the centuries that are marked by us 'geeks' shaping so much of the way humans live.

The key to the whole transformation brought about by science was to look *outward* at physical evidence to answer a question, rather than looking *back* at tradition or *inward* to revelation to settle disputes.

With his telescope, Galileo quickly demonstrated that the moon was not a perfect, spherical body as the ancients had thought, a nice smooth bowling ball left over from the first moments of creation. It was a complex planetoid, with light-colored highlands, dark lowlands, and major craters crisscrossed by little craters—everything that made it "imperfect" to people of 1609.

It was as if perfection itself has fallen to the ground, and old traditions died with the crash.

More remarkably still, even the sun was found to be imperfect. The brightest, seemingly most pure body in the sky was round enough, but it had moving blemishes, spots right on its face!

The foundations of the ancient way of looking at the universe crumbled, never to be revived again.

Galileo also soon made observations of Venus, the planet so bright it's easy to pick out in the sky some evenings before the sun has fully set. He showed that Venus waxes and wanes a bit like the moon as seen from the Earth. That was a crucial observation about the solar system, because tradition and the Church taught that all bodies in the sky revolve around Earth. Galileo's diagrams of how Venus waxed and waned like the moon made it clear that Venus was not revolving around Earth, but was, indeed, revolving around the sun.

Galileo also looked at Jupiter in the night sky—which his telescope revealed had bodies orbiting around it. Again, the visual evidence made it clear that the Earth was not the center of the solar system.

Galileo, to be sure, stood on the shoulders of others. Copernicus was one obvious giant who had come before him. Good work always depends on a network of earlier efforts. But it's also true a crucial corner was turned in 1609, and nothing has been the same since.

Dynamite Brains

People are unusual animals. We spend a lot of time watching mediocre TV, but we also describe our thoughts in elegant and written language. We pour beer on our heads at football games, but we also study the whole history of life on Earth as it is preserved in fossils.

It's not our biceps that make us special, but our thoughts. What makes our thinking so complex, able to soar with the poets and solve problems with the engineers?

In the 1800s, a number of scientists spent considerable effort trying to determine a biological basis that they assumed made some people "smarter" and "more civilized" than others. We can learn something of vital importance about science and scientists from the history of their projects.

Craniology, as the science of head size and brain structure was called, had two basic ideas: bigger brains were smarter brains, and any structures that were toward the front of the brain were more developed in smarter or more civilized people.

The first major craniologist was an American named Samuel George Morton. Skulls give a good measure of brain volume, and they keep much better on a shelf over time than brains do. So Morton collected over 1,000 skulls from all over the world. He examined them in the best scientific spirit of the day. In one of his techniques, he poured lead shot into skulls, then poured the shot out to measure its volume.

Morton's published papers argued that the skulls of whites were larger than those from subjects with red, yellow, brown, or black skin. Women, Morton simply assumed, were so intellectually ungifted he barely mentioned the fact that their brains were smallest of all in his measurements.

Morton was confident enough of his work he published his raw data—the measurement of each skull or set of skulls. And that made possible a "double check" of his work more than a century later.

Stephen Jay Gould, the famous professor of geology and paleontology at Harvard University, for whom I once worked as a teaching assistant, plowed through all of Morton's measurements and the arithmetic he published. (You can find a forceful treatment of the story in Gould's *The Mismeasure of Man*, available at libraries everywhere in the nation and via second-hand bookstores.) Gould found that Morton—and other scientists who followed in his footsteps such as Robert B. Bean—consciously or unconsciously fudged their data to support the conclusions they fully expected to find.

Just as one example, women of every race are, on average, of smaller stature than men of that race. I'm a woman 5 foot 6 inches tall. My friend Peter is of my race, and he is 6 foot 6 inches tall. Guess who has a larger hat size, Peter or me?

Here's the kicker: Morton's tables sometimes had only women's skulls for races he viewed as inferior. He didn't view that as a problem. But for "Englishmen" (whom he viewed as a category worthy of separate record-keeping from other whites), he measured only male skulls—no female skulls allowed in the data!

Morton cooked the books in many other ways, too. His arithmetic is wrong in several places where he computed averages—and his errors elevate the brain size of white males and depress the brain size of non-whites.

It's easy to mock craniology now, but it was a serious pursuit in its day. The point of remembering it is that we should always be on guard for the errors any professional can make due to the climate of the times. Science is a profession pursued by human beings, not objective robots.

As readers of this column know well, I'm a committed believer in science. All around us the evidence is clear that scientists make great discoveries each day. But it's worth remembering that we scientists also make some real mistakes.

After all, if craniologists had thought clearly for even a moment, they would have abandoned their work. Brain size cannot explain much, if anything, about intelligence. Neanderthals had modestly bigger brains than we do. And what about elephants and blue whales?

As my fourth grade teacher used to say, "Put your thinking caps on!"

Future Looks Brighter than the News

I've seen the future of American science and engineering. And, in my humble opinion, it looks very bright.

From time to time the media tell us that American education simply isn't working. Reports can make it seem that public schools—and universities, too—are wasteful, dysfunctional, and produce students who can neither read nor write, let alone do science and math.

But I work at a large state university and I see little evidence of those claims. Let me tell you what I do know about, what I see first hand.

Recently students majoring in all the sciences and engineering here at Washington State University presented the results of their research to both faculty and industry representatives from outside the Ivory Tower. Yes, I said that the undergraduate students—some of them 19 years old—presented the results of their own research to faculty, staff, their peers, and industry representatives.

The next generation of nerds is more involved in research work than any I've known to this date. They don't just sit passively in class taking notes, but broaden their horizons and deepen their minds by pursuing real research work in faculty labs on questions ranging from the characteristics of different types of steel to approaches for characterizing and combating onion rot.

It's quite a sight to see a crowd of 300 young people doing the same basic exchanges—explanation and a bit of argument—that happens at national professional meetings where 55-year-olds explain and defend their research to peers. The ballroom where we held this event was crowded with people, the noise level high due to all the intense conversations. It's all obviously good practice for the students, and the research being done by them simply puts to shame the education some of us old-time gals had years back. And that means the next generation can benefit society sooner in their professional careers.

The basic quality of all American universities is shown by the fact that many students from overseas still come here for their education. They vote with their feet—and some of them stay to contribute to our society as scientists, engineers, doctors, and more.

One of the areas in which Americans have long excelled is basic research in science and engineering. We have been the go-to nation for computing technology and software, obviously, but also for a host of other technical fields like biotech science that allows us to change DNA for useful purposes and geological engineering techniques that allow us to explore for and extract oil from the deep Earth with the least impact on the environment.

Recent years, of course, have been tough. It's easy to be weary of the stories on the nightly news, to be afraid of the economic stresses our nation is going through. But I honestly believe the kids of today (if I may be forgiven for calling college students 'kids') have all that it takes and more to compete globally in technical fields. They are being taught to think and to do, not to recite—and that's to our great advantage as a society.

True, American universities are under major financial stresses. But I know first hand that American universities are still doing good work, day in and day out. The results are clear in the research I being done not just by faculty and graduate students, but by teenagers who are throwing themselves into technical life.

While I was helping with the event, two parents and their daughter came up to me. The daughter is admitted to attend the university next fall and she has interests in majoring in a technical field. It's fair to say they were blown away by the event, because it made clear both what a great education can be had at a public university and how serious some people in the upcoming generation are about their intellectual and professional lives.

I don't really understand economics, but I do know contributions to economic competitiveness by scientists and engineers are very real. And we here in the Ivory Tower are raising a good crop, indeed, to replace ourselves—so the nation can climb still higher.

Pioneers of Another Sort

Sometimes it pays to spend ten years in detention. Not that a person would ever want that to happen, but if it did—could you put the time to good use?

That's a question I've asked myself. I've also sometimes asked my students exactly the same thing. The value of a good high school or college education, I say to them, is that it should give you the tools to use time like that well. What would you do with it?

One thousand years ago an Arab man named Ibn al-Haytham found himself under house arrest in Cairo. That far back ago in time, we don't know much of the specifics of Ibn al-Haytham's life. But we do know he was a towering giant of an intellectual in his day.

If you give a thinking person ten years to think, don't be surprised if there are some powerful results. In Ibn al-Haytham's case, a good argument can be made that the ten-year gap in personal freedom was quickly followed by the release of his major book on optics. That book was pivotal to our lives today, because optics was hardly the only issue it addressed.

In the ancient world—more than 1,000 years before Ibn al-Haytham's own life—Greek philosophers had two main theories of vision. One theory (advanced by Ptolemy and Euclid) was that "vision rays" left the eye and went out to objects around us in the world. The other was put forward by Aristotle. The great philosopher had argued a "form" of some sort comes from an object in the world around you and enters your eye so you can see it.

Ibn al-Haytham pointed out, first, that not all the ancient Greek authorities could be right, since they followed two contradictory ideas on the subject. Then he noted that we don't have vision unless there is light around us: either light from the object we are seeing (like a lamp) or light rays from reflected light (like sunlight during the day). So light, first and foremost, is what we need to understand in order to better understand vision.

Using only logic like this and a few simple experimental materials—a pinhole in a curtain or a hollow straight tube—Ibn al-Haytham went on to deduce a great deal about modern optics. Here's the basics: Light rays travel in straight lines. Light on flat mirrors is reflected in one set of ways, and on curved mirrors in others. Light is refracted (bent) when it moves from air to water or back again.

Most importantly of all, Ibn al Haytham did all this good work using experiments and observations, writing out for his readers what they could do to show themselves *the same evidence* he had seen. This meant they would reach the same conclusions as he had, not because they should blindly accept his ideas, but because they could recreate the same data he had observed.

Putting forward that core of the empirical spirit is not bad for 10 years of work spent under nice conditions. For 10 years in detention, it's really a remarkable feat.

Two hundred years passed after the death of the Arab scholar before a Christian monk took up a translated volume of the work and saw its value. Roger Bacon was our hero's name. He was not Francis Bacon—there are two Bacons rattling around in history. This, we could say, is "the other Bacon." Roger Bacon repeated some of Ibn al-Haytham's experiments—but more importantly he also endorsed for the Christian audience this new method of gaining new knowledge about the natural world. Experiments and testing of physical facts, Bacon argued, were the most productive ways to learn about the physical world around us. Others around Bacon were soon on board with the program, and Medieval Europe began to have at least an inkling of the modern, scientific method.

The reason science and engineering have been able to progress so much in our lifetimes is that the method of running experiments and testing results is enormously successful. But in the old world, it was far from clear that this approach would lead to the soundest results.

We owe Ibn al-Haytham and Roger Bacon a lot, not just for their good work on optics, but for recognizing the power of the scientific method that has given us so much today.

SECTION 9
ECLECTIC TOPICS

T he columns I've written as the Rock Doc cover a wide range of subjects. Some of the essays relate to matters that don't fit into any firm categories—perhaps like the Rock Doc's own rather chaotic thinking.

Eclectic Topics

Fetching for a Living

My brown mutt and I went for a walk recently on an old railroad grade at the edge of a ghost town where we often stroll. Buster Brown is a Lab mix, with an emphasis on the mix.

Buster likes to visit the ghost town in part because it still has one occupied house, and friends of mine live there with their chickens, cat, and dog. The resident dog is a small, insanely intense cattle dog mix. He dedicates his considerable energies to retrieving pine cones kicked along the road.

While Buster is willing to retrieve sticks thrown into water, he won't play that game for more than 10 minutes. But I can't tire out his country friend, who will chase pinecones as long as you are willing to kick them. The little cattle dog has a deep and unshakable concentration for his "work."

Humans have bred that kind of intensity into some dogs. And science is now revealing that along with focused energy, some dogs have more intelligence than we ever really understood. Here's the story.

A border collie in South Carolina named Chaser was adopted by a research psychologist after the man retired and had free time. John W. Pilley took Chaser home when she was eight weeks of age and started intensive, five hours per day training of the young pup. For three years the dog worked with Pilley and a few others, learning a variety of commands and behaviors through the process of game playing, with nothing more than verbal praise as a reward for a job well done.

Only a border collie would love so much schooling, but Chaser flourished in her new life.

What's impressive is how much Pilley was able to teach the collie to ultimately do. Pilley acquired second-hand stuffed toys and went to work teaching Chaser to fetch individual toys by name. As the total number of toys Chaser knew edged into the hundreds, Pilley had to write their names on them with a marker so he could remember them all. In the end, Chaser learned over 1,000 proper names for the toys and could reliably fetch them from another room or a different part of the yard and bring them back to Pilley.

She also learned what you could term "group names." For example, some of the toys were Frisbees, and some were cloth disks you throw in the same way. Chaser learned specific names for such toys, but also the general category "Frisbee." This means she will fetch a "Frisbee"—whatever one may be available—if you ask her for a "Frisbee." But she'll also bring you one particular disk if you ask for it by its specific name.

What's even more impressive is how Chaser responds to situations that might overwhelm a 4-year-old human. Let's say you put a new toy among the ones Chaser knows—a toy she has never seen before. Then you ask her to fetch and give her a specific name she's never heard before. Chaser goes to her toys, looks them all over, and apparently thinks pretty deeply about her work. She then selects the toy she's never seen and doesn't know. This is called "learning by exclusion" and is the mark of some serious thinking in her furry noggin.

It's not clear that border collies really need to be so smart. The commands they follow for a shepherd rounding up sheep are not so complex as learning by exclusion, nor do the collies need to know thousands of individual sheep names. But, somehow, it seems that in the process of breeding border collies for their indefatigable willingness to work for us around the clock, we have created truly thinking canines.

One thing seems clear. Chaser looks awfully smart to us now not because she was a really special border collie pup, but because she fell into an extremely rich learning environment.

Some would say my Buster Brown is not in the same mental league as Chaser. But if Buster could speak he might point out he gets his kibbles every day just like she does, and without slavish devotion to his work.

Who really is the smarter dog?

Feeling Cautious about Precaution

Each day I take an anti-inflammatory medication that makes it possible for me to move around despite my joint problems. The drug helps beat down my arthritis pain, and I'm grateful for that. But it's also a fact that the medication increases the chance I'll hemorrhage internally. I choose to accept that risk both because leading a completely sedentary life doesn't appeal to me and because there are real, physical dangers in never getting up from a chair.

The way I figure it, there are risks in taking the medication (and living as I wish) and risks in not taking it (and facing the dangers of inactivity). What's important for me to acknowledge is that it's not a choice of perfect safety versus great risk, but of one set of risks versus another.

In recent years you may have heard of the "precautionary principle," an idea meant to help us choose wisely in situations with risks and hazards. One way of stating the idea is that if an activity (like applying an herbicide to a crop in the field) could harm human health or the environment, precaution is the best way to proceed even if the exact risks are not yet fully known. "Better safe than sorry" sums up the notion at a gut level. And, beyond that, the precautionary principle as it's usually put forward states that doing something bears the burden of proof, not refraining from doing anything. So spraying the field bears the burden of proving itself a safe course of action, but watching weeds grow does not.

In my own medical case, we could apply the precautionary principle like this. If I do nothing, I run the real, medical risks of being sedentary. But by taking anti-inflammatory medication every day, I run the risks of side effects so serious I end up in the ER with my life at stake. The precautionary principle seems to advise that I should not take the drug that will help keep me active, because that overt action exposes me to real and significant risks. It's the activity (taking the medication) that bears the burden of proof with respect to safety, not letting nature take her course within my joints.

I'm for all of us deciding on our own what medications we will take. But when it comes to some bigger questions—like what herbicides should be available in the United States—what I think starts to impinge on your life and vice versa. All herbicides in this country go through rigorous testing before they are approved for market. Yet it's also true that Herbicide X may pose some small level of risk when it's employed—either to field workers, consumers of a crop, or wildlife living downstream of a field treated with the herbicide.

The precautionary principle, at least as I've seen it employed in environmental questions, is quite one-sided. The risk of using Herbicide X is evaluated against perfect and idealized safety, not against concrete options in the real world.

There is some risk in using a chemical on a crop. But there is also a real risk in doing nothing. That risk includes lower crop yield, which could easily entail more people having less to eat, as well as more farmers in bankruptcy court.

Let's go back to the analogy with my medication. I'd like someone to wave a magic wand over me and take away all traces of arthritis in my body. That would be peachy, especially if it could be done at no risk to any of us. But the realistic choices are that I take anti-inflammatory medication so I can move around or that I don't take medication and live a wholly sedentary life. Both options entail risks. Unless I can admit that, I don't see how I can begin to intelligently choose what I want to do.

And unless we can discuss real risks on both sides of environmental issues, I fear we will remain locked in discussions laced with passion rather than meaningful dialogue—let alone thoughtful decision-making.

Doing the 180 Pivot

I remember well when Barbara Mandrell and George Jones sang a fine country and western song about changes in public perspective. For those of you too young to remember, the lyrics of the chorus were:

"I was country when country wasn't cool/I was country from my hat down to my boots/I still act and look the same/what you see ain't nothin' new/I was country when country wasn't cool."

The song rocketed up the charts when the country twang had become the fashion in popular music—a testimony to how public opinion can turn on a dime because country music hasn't been popular outside its niche before nor since.

Here's another example of how public perspectives can radically change. It wasn't that long ago that most environmentalists viewed the wood-products industry as the enemy. The image of lumberjacks cutting down trees seemed to many like a picture of rapacious wolves gobbling up innocent bunny rabbits.

But opinions have changed about that, too.

"I think some of my green friends in the old days thought I was a 'lumber-Nazi,'" Dr. Karl Englund said to me recently.

Englund is an engineering researcher in wood products and their uses at Washington State University.

"But there's been a real 180 degree turn. Now people understand that wood products are a renewable and green resource," he continued.

Englund spoke as he gave me a tour of the research and testing facility where he works.

"You can think of wood as solid carbon dioxide, exactly what we need to take carbon dioxide out of the air and put it into 'sinks' where it will stay for a long time," he said while knocking—of course—on wood.

To be sure, there's nothing new about lumber. But what is new are the materials that people like Englund dream up and produce out of scrap wood, saw dust, waste plastic, and other materials of little value.

Laminated veneer lumber, or LVL, was one of the first products to come out of wood engineering labors. LVL is made of thin but large layers of wood that require a full-sized log. They are then pressed and glued together to form material that's stronger and more uniform than lumber. Oriented strand board, or OSB, is also useful because it's made of smaller strips of wood in layers that are oriented in alternating directions.

Another, newer material that's a topic of research is formed from short strips of wood that are heated and pressed into 3-D forms that create what engineers call a "complex geometric shape."

I'd call it a wood waffle. It's interesting stuff just to toss around.

Now think of such a waffle that has a thin veneer of wood on its top and bottom. What you have is a material that's lightweight, strong, and traps air in its pockets, making it naturally insulating.

That's good engineering all around, the type of thing Englund and his colleagues are researching for possible transmission to the commercial sector.

But the story gets better from a green point of view.

Smaller pieces of waste wood can be ground up to material similar in size to whole-wheat flour. When combined in a giant extruder with plastic—including recycled plastics that otherwise would be waste—the fine wood fiber particles form the strength component of products like composite lumber decking. The material doesn't rot outdoors because the plastic seals the wood away from moisture. Add some coloring agents, and you have decking that looks fine and will never sink a splinter into your bare feet.

Good engineering isn't limited to wood products. When material is ground up to small sizes, wheat straw, rice straw, and other agricultural by-products can be used in some basically similar ways.

Englund has a refreshing perspective on his engineering work.

"I'm a garbage guy," he said. "The point is to use people's waste."

Englund may work mainly within the wood-products industry, but it's clear to me he's really green. And he always had been.

Unlike public opinion, some things are a constant.

And yes, I'm still listenin' to country music—sometimes in my best western hat.

The '87 Truck

I came to a sharp fork in the deeply rutted road of my life this fall. I had to decide if I would continue to limp around on Saturdays in my beloved but inefficient '87 pickup, or sell it off to some poor soul in more need of it than I.

My eight cylinder American-made truck has a relatively small engine in it, the most petite offered in its day. Still, you can feel the engine torque the body of the truck when you turn it on. Perhaps that's why it gets only about a dozen miles to the gallon, and that's at 50 mph with a strong tail wind. If I'm towing anything, or have a heavy load in the truck bed, the miles per gallon figure crashes into the single digits. In short, my fine truck is not what you'd call fuel efficient.

But a rural geologist needs a truck. It's part of the image, isn't it?

Still, lately I've been strongly tempted to sell the thing off. The truck costs money in gas and oil. It requires funds to insure. New studded snow tires add to the fun. The truck is old enough it breaks down, including in the middle of the road, leading to towing bills and major charges from my mechanic. (I suspect my pickup pays his mortgage, but perhaps it only seems that way to me.)

There's always the dream of a new, more fuel-efficient truck. There are models that actually shut down some of their cylinders on the highway when they are not needed, helping to boost miles-per-gallon figures. Or I could switch to a smaller truck entirely, making efficiency gains hand over fist. And, of course, the new truck could be a glaring yellow with red flames painted down the sides. What could be better?

But perhaps Americans have learned something in this recession. Maybe rather than allowing myself to be seduced by an expensive new truck I'd seldom use, I'm better off sticking with what I've got. My ride isn't swank. The cab is smelly, a mix of the aroma of wet dog, decaying cushions, and deep-seated mold. The engine runs a touch rough sometimes. It burns a bit of oil. I don't get great gas mileage, but I don't put many miles on the truck. And I can operate and fix my truck for far less than the price of the annual interest of a new truck, even one without flames painted on the sides.

It may be corny to say it. But most citizens of this fine republic have had to reexamine at least some of our financial priorities in recent years. I know it's felt that way to me.

I could, in truth, live without a truck at all. But it's convenient to have one for taking loads of yard waste to the dump, for lending to young and strong folks moving from one home to another, for towing utility trailers, and for taking a load of household goods (where does all the stuff come from?) to the annual church rummage sale.

So I'm going to keep my old truck. It's not always convenient to use because it does break down. On the other hand, I can think of it as an adventure every time I drive, something no new truck offers an owner. Maybe that's the kind of out-of-the-box thinking we need to hang onto during this long crawl upward out of the deep recession into which we fell.

While I'll stick with what I've got, I like to look at modern trucks, mostly because they are marvels of engineering. And when I get too worried about the economy, I remind myself there is still a powerhouse of engineering research in this large and diverse country of ours. Everything from modern medicine to electronics to agricultural engineering to research in genetics is rapidly gaining ground. Perhaps with some renewed personal values about what's really important, and our bedrock national advantages in sciences and engineering, the United States will see progress in the coming years in many important respects.

May it be so.

Let There Be Light

We all make a thousand choices about the energy we use each day and we pay for each decision we make.

Some choices have big consequences, like living close to work or putting a woodstove into the living room. One smaller choice consumers have been used to is deciding what kind of light bulb to buy, a matter that can affect a percentage of our electric bill.

The spiral-shaped compact fluorescent bulbs use less electricity than the incandescent bulbs we grew up with. That saves you money on your power bill, and it also means less carbon dioxide is added to the planet's atmosphere from power plants. The new bulbs recently caught the attention of Congress. The upshot is that federal law pushes us to switch to the new bulbs starting in 2012. But as you consider your light bulb options, there are a couple of things you'll want to know.

At least as they've been manufactured so far, compact fluorescent bulbs create light by making mercury gas within them fluoresce. As older readers will remember, forty years ago thermometers had mercury in them. Mercury was the silver liquid that rose and fell in the thermometer, registering the temperature of a feverish child.

Mercury is a well-known neurotoxin. In fact, there's an old phrase that reminds us of that fact. Mercury was once used in the processing of fur pelts that made hats, so people who worked in the hat-making industry were exposed to a lot of mercury. And it was mercury's clear and strong effects on workers that led to the phrase "mad as a hatter," as well as to the character you'll recall as the Mad Hatter in *Alice in Wonderland*.

If you break a compact fluorescent bulb, either when you drop it or toss it into a recycling bin, mercury vapor wafts into the air. Here are the federal recommendations for what to do if you break a bulb indoors.

1. Open a window and evacuate the room "for 15 minutes or more."
2. On a hard surface: using gloves, scoop up fragments of the bulb and powder and put into a sealed plastic bag; wipe the area with damp paper towels and put them into the bag.
3. On a carpet: using gloves, scoop up fragments of the bulb and powder and put into a sealed plastic bag; press sticky tape, such as duct tape, into the carpet to remove more material, and add the tape to the bag; vacuum the area; remove the vacuum bag from the device and put it

in the sealed bag; wipe down the vacuum with damp clothes and put them into the sealed bag.

4. Put your sealed bag into another bag, creating a doubly-sealed bag.

5. If your local laws allow it, put the materials into the trash.

6. Wash your hands with soap and water.

On the upside, a compact fluorescent bulb contains only about 10-20 milligrams of mercury. For reference, an old thermometer contains 500 to 2,500 milligrams depending on its internal size. (Note: some of us geezers broke the old-style thermometers when we were young, played with the mercury on our palms, and lived to tell the tale. But maybe that experience accounts for our peculiar thoughts?)

Although the compact fluorescents are supposed to be recycled—without breakage—many of the bulbs in fact head to landfills where they are crushed. Although the amount of mercury in each bulb is small, the number of such bulbs in the United States is now staggering and still headed north. Another concern is that bulbs are almost all manufactured in China, where industrial hygiene standards are not always the best, and workers may become more like the hatters of old than consumers shall ever be.

The good news is that the next generation of lights are on the way in the form of Light Emitting Diodes (LEDs). The tiny bulbs can be even more efficient than compact fluorescents.

Here's hoping LEDs can help us reduce our electric bills without promoting effects worthy of Alice in Wonderland.

Summary Vacations that Teach a Lot

My finest memory from childhood is sailing a kayak my clever brother made out of plywood and canvas. The sailing adventure was on a lake in Glacier National Park. The wind was good, my beloved dog was tucked between my knees for ballast, and I scooted over the water like a rocket—or so it seemed to an 11-year-old. I have a black and white photo of that event (the world was black and white in those days), and I cherish it greatly even though it's faded.

My family's visits to Glacier National Park when I was a kid may have had a lot to do with my eventually studying geology in college. And when I read geologic literature that's coming out more recently, I find I still learn a great deal from part of our national heritage we've set aside for future generations. Indeed, what geologists have learned in Glacier National is even helping them interpret current tectonic events in the Himalayas and the Andes.

Let's start at the beginning. The rocks of Glacier National are ancient, going back over a billion years. And they are well preserved. They are sedimentary and show us simple fossils from that ancient time. The fossils are colonial clumps, if you will, of single-celled creatures that lived in the ancient seas. We call them "stromatolites," and they and the chemistry of the rocks in which we find them help scientists understand the climate and atmosphere of that ancient time.

Long after the stromatolites, the rocks were deformed by the great crunch of ancient tectonic plates. This formed an enormous fault called the Lewis Overthrust that moved rocks up to 50 miles laterally. It was all part of the events that helped make the ancestral Rocky Mountains, similar to events happening in major mountain chains in South America and Asia today.

Now fast-forward with me from the ancient part of Earth history to the much more recent Ice Age. This was the period when enormous glaciers covered much of North America. It's my favorite part of history, when woolly mammoths and saber tooth tigers roamed our land.

Glaciers are superb agents of erosion and they made deep, majestic valleys in the ancient rocks now in the park. It's the steep sides of the valleys that allow us to see the stromatolites and all the other features of the ancient rocks. And it's the deep, majestic valleys that take your breath away and keep people coming back to Glacier National year after year.

Even a casual visitor to America's various parks and gardens can learn a lot about Earth history and the Ice Age. In Manhattan's Central Park, for example, some of the rocks show the effects of Ice Age glaciation. Giant blocks of rock moved many miles by glaciers, called "erratics" because they don't belong in the area, are littered around the park. And the bedrock shows deep striations or grooves carved by the rocks trapped in the glacier as it moved. So even in urban areas, the natural world can speak to us if we are in the frame of mind to listen.

By the way, besides sailing that kayak, another fine event in Glacier National stands out in my mind. Near streams and in the lowlands there were plenty of huckleberries to pick. What can be better than foraging for your own food on summer vacation when you are a town kid used to hot lunches at schools? I owe my parents a lot, I think, in giving me those early memories.

Now that I'm older I realize one reason we made road and camping trips when I was a kid was that spending time away from home in that mode doesn't cost as much as some other means of having a getaway. But the fact that local park visits or camping trips are low budget doesn't mean they are low value.

Seize the day come summer, in our National Parks or just someplace closer to home. Picking up rocks or hunting fossils doesn't cost much, and the soul you enrich may be your own.

Suggested Reading

The following books, journals, and websites allow readers to further explore science in lay terms.

Alley, Richard B. 2000. *The Two-Mile Time Machine: Ice Cores, Abrupt Climate Change, and Our Future*, Princeton University Press. This book is written by a major player in the field of climate research. He worked on the Greenland ice core project, which taught us much about abrupt climate change.

Alley, Richard B. 2011. *Earth: The Operators' Manual*, Norton. This is Alley's most recent book. It's a long tour of several related topics, with digressions and humor thrown in. Alley argues that carbon dioxide emissions from people are causing global warming. He also discusses energy policy and what he sees as the innovations needed in the energy realm. The book is a companion to a PBS documentary.

Bryson, Bill. 2003. *A Short History of Nearly Everything*, Random House. Bryson is a bestselling author who has explored many different topics in a variety of books over the years. In this thick volume he tells the tale of the history of science and, in the process, draws the reader into many technical details.

Chase, Alston. 1995. *In a Dark Wood: The Fight Over Forests and the Myths of Nature*, Houghton Mifflin. Chase writes well about complicated topics in this forceful volume, which the Rock Doc particularly recommends. The book discusses the history and the effects of the conservation and environmental movements in the United States.

Cox, John D. 2005. *Climate Crash: Abrupt Climate Change and What It Means for Our Future*, Joseph Henry Press. This is a gem of a book that takes the reader quickly through the core of natural climate change, including the evidence that climate changes quickly and takes down human societies when it does so.

Diamond, Jared M. 2005. *Collapse: How Societies Choose to Fail or Succeed*, Penguin. All societies and civilizations encounter stress, including from the natural world. This bestseller takes the reader on a tour of how past societies have adapted—or disappeared.

Fagan, Brian. 2000. *The Little Ice Age: How Climate Made History, 1300–1850*, Basic Books. The warm climate enjoyed by Europe in the high Middle Ages gave way abruptly in the 1300s to cold years with wet summers. The new, grim period—misleadingly called "the Little Ice Age" by historians—did not end until 1850, when our current balmier times began. Fagan is a historian who has written two books about climate's impact on human history; see the next entry for his other book in this arena.

Fagan, Brian. 2008. *The Great Warming: Climate Change and the Rise and Fall of Civilizations*, Bloomsbury Press. The Middle Ages were characterized in Europe by warm summers with relatively good harvests. This book, like Fagan's other one in this list, explains just how important the weather was for the development of European civilization.

Fenton, Carroll Lane, and Mildred Adams Fenton. 1952. *Giants of Geology*, Doubleday. This is an old book, likely available to the reader through interlibrary loan or second hand bookstores. It's an easy read, exploring major contributions from a number of great geologists of earlier times.

genographic.nationalgeographic.com. This website contains information on a large genealogical study, along with instructions on how to participate and learn more about your genetic history.

Gore, Al. 1992. *Earth in the Balance: Ecology and the Human Spirit*, Rodale. The former vice president sets out his main environmental agenda in this book. This framework has remained with Gore in later years, including in his documentary *An Inconvenient Truth*.

Gore, Al. 2009. *Our Choice: A Plan to Solve the Climate Crisis*, Rodale. Former Vice President Gore has written several books about environmental topics, the political landscape, and climate change; this volume is one of his most current offerings.

Krajick, Kevin. 2001. *Barren Lands: An Epic Search for Diamonds in the North American Arctic*, Times Books. Exploration geologists are a different breed of human, often trading an easy life for hardship and danger in the back-of-beyond. This book takes the reader on a long but breathless trip with the seat-of-the-pants work that led to the discovery of a cornucopia of diamond claims in northern Canada. If you think geology is romantic, this book will give you a gripping read over several long, winter evenings.

Luire, Edward. 1960. *Louis Agassiz: A Life in Science*, University of Chicago Press. This is the definitive book about the man who first argued for the existence of the Ice Age, a period of time that determined the landscape of Europe and the northern United States, as well as shaping animal species like the woolly mammoth.

Macdougall, Doug. 2004. *Frozen Earth: The Once and Future Story of Ice Ages*, University of California Press. This is a readable introduction to the Pleistocene Epoch, informally known as the Ice Age.

Muller, Richard A. 2008. *Physics for Future Presidents: The Science behind the Headlines*, Norton. Muller is a physics professor at the University of California, Berkeley. He has taught a freshman level course for non-science majors; this book is an outgrowth of that effort. Starting with the events of September 11, 2001, and running through energy policy and global warming, the book gives the reader a scientific tour that includes valuable information not presented elsewhere at this level. If you have interests in our national energy policy, read this book.

Natural History Magazine. This magazine is designed for the interested layman who has curiosity about a broad range of topics related to science. It's available by subscription and at many public libraries. Its website is naturalhistorymag.com.

Peters, E. Kirsten. 1996. *No Stone Unturned: Reasoning about Rocks and Fossils*, W.H. Freeman and Co. This small volume by the Rock Doc explores the history and philosophy of science using geology as the vehicle for discussion.

Peters, E. Kirsten. 2012. *An Inconvenient Climate*, Prometheus Books. This book escorts the reader through the history of how geologists deduced many natural changes in Earth's climate—and what that story means for us now.

Pickens, T. Boone. 2008. *The First Billion is the Hardest: Reflections on a Life of Comebacks and America's Energy Future*, Crown Business. Pickens is a geologist who has made—and lost—many millions of dollars during the course of a long career in the energy field. In this book he gives autobiographical details of his life and also argues for his energy plan for the nation. As Pickens sees it, we should save our natural gas for transportation needs and use as much wind power as we can to help power the national electrical grid.

Pollan, Michael. 2006. *The Omnivore's Dilemma: A Natural History of Four Meals*, Penguin Press. Pollan's book has as its subtext the criticism of modern agriculture, in particular the way we grow and use corn in this country. The book explores how small farmers can produce many different foodstuffs from the same plot of land.

redcross.org. The American Red Cross has an extensive website with a lot of information. Look especially for their suggestions on preparing a family emergency preparedness kit.

Ruddiman, William F. 2005. *Plows, Plagues, and Petroleum: How Humans Took Control of Climate*, Princeton University Press. Ruddiman is a retired climate scientist who, in this small but valuable book, argues that early agriculture helped shape climate long before the invention of the steam engine. If you want an alternative view of the relationship between people and climate, this is a great place to start.

Rudwick, Martin J.S. 1972. *The Meaning of Fossils: Episodes in the History of Paleontology*, University of Chicago Press. If extinct species and the work of paleontology is up your alley, this book offers a serious overview of how natural scientists first looked at fossils and their significance to Earth's history.

Schneider, Reto U. 2008. *The Mad Science Book: 100 Amazing Experiments from the History of Science*, translated by Peter Lewis, Fall River Press. This volume explains small bits of scientific research that are weird, crazy, or humorous. It's helped along by an abundance of cartoons and photos.

science360.gov. This website, which dubs itself the Knowledge Network, is updated often and has stories on a wide variety of topics related to research science. If you have access to high-speed internet, check out this site.

sciencenews.org. This website is designed to help promote public awareness and understanding of scientific research projects.

Snyder, Carl H. 2003. *The Extraordinary Chemistry of Ordinary Things*, fourth edition, John Wiley and Sons. If you want to read an actual chemistry textbook, but one that tries to meet you halfway with abundant examples of chemicals in your daily life, this is the book for you.

Weart, Spencer R. 2003. *The Discovery of Global Warming*, Harvard University Press. This is a dense but readable history of how scientists came to understand the many complexities of climate and began to see that human activities may be moving us to a much warmer future.

Webb, George Earnest. 1983. *Tree Rings and Telescopes: The Scientific Career of A.E. Douglass*, University of Arizona Press. This book takes the reader through the life of the man who almost single-handedly established tree-ring research as a valuable part of both climatology and archeology.

About the Author

D r. Elsa Kirsten Peters grew up one block from the edge of the Washington State University campus, where her father was a faculty member. As a child, she fell in love with the Ice Age. As a young woman, she majored in geology at Princeton University and earned her doctorate in the same subject from Harvard. She has published technical abstracts and articles in her field of expertise, as well as essays about science, religion, and the challenges of teaching college students to write. She has taught geology and interdisciplinary science classes at Washington State University, where she is a faculty member. She is the author of two textbooks in geological science and has been a freelance editor for college-level materials in oceanography, environmental science, and geology. She is the author of *An Inconvenient Climate*, in press with Prometheus Books, which is a popular treatment of how geologists learned about natural climate change. This volume is her seventh published book.

Be sure to keep up with the Rock Doc's continuing adventures in her weekly newspaper column. If your local newspaper doesn't carry Rock Doc, encourage them to add it to their lineup. As always, you can access all of Rock Doc's published columns for free at rockdoc.wsu.edu.

Planet Rock Doc includes many of the Rock Doc's favorite columns, plus several columns published for the first time in this volume: "Breaking a Rock to Get at Oil and Gas," "Too Much Exercise," "Twins, DNA, and Mental Illness," "Fetching for a Living," and "Feeling Cautious about Precaution."